INVESTMENT PROPERTY VALUATION TODAY

by

TIM HAVARD

2004

EG
BOOKS

A division of Reed Business Information
ESTATES GAZETTE
151 WARDOUR STREET, LONDON W1F 8BN

First published in 2000 by Chandos Publishing (Oxford) Limited
This reformatted edition published by Estates Gazette in September 2002

Reprinted 2004

ISBN 0 7282 0393 6

Contents

Preface

The philosophy behind this book is based on three related observations about the current state of investment property valuation:

- The valuation of investment (ie income-producing) property has been complicated by radical changes in market conditions and lease structures over the last decade.
- There are a number of valuation books for students of property valuation but very few that are aimed at practising valuers and these generally do not keep pace with current market conditions.
- There have been a number of developments in various theoretical aspects of valuation in the academic world that would be useful to practising valuers if they were presented in a more accessible way.

At present valuation surveyors wanting to extend their knowledge are forced either to tackle articles in academic journals (often written for other academics), or else attend expensive and time-consuming continuing professional development (CPD) events.

This book is an attempt to reconcile the three observations. The book is divided into two parts. In the first, the idea is to present an overview of the main techniques available to the valuer, both familiar and unfamiliar, giving a simple description of the principals of each. This includes a critical appraisal of the qualities of each method. If readers require more in-depth coverage, the bibliography and references will lead them to recent source material. In the second part, real-world problems are examined with each of the methods of valuation used to illustrate the strengths and weaknesses of each approach.

The Author

Tim Havard has a varied background in construction and property. Following the completion of an undergraduate degree in prehistory and archaeology he was trained as a quantity surveyor with a construction firm in Cumbria. On completion of training he obtained a postgraduate qualification in land economics from the University of Aberdeen. He then qualified as a general practice surveyor and worked in private practice in the south-east of England and Scotland specialising in landlord and tenant, commercial valuation and investment work. In 1990 he joined the life assurance company Scottish Provident as an investment surveyor, managing a portfolio of commercial properties and being responsible for a number of development projects. In 1992 he took up a lectureship with the University of Technology, Sydney, Australia. Following his return to the UK he joined UMIST in September 1995 where he taught property investment and development at Masters level.

He has given papers on valuation and property investment at a number of international property conferences and has had articles published in leading academic journals. He has recently completed his PhD that investigated links between valuer behaviour and valuation variance.

Chapter 1

Introduction: The Market Framework

UK valuers have a reputation for conservatism (with a small 'c'), disliking innovation and change in the methods that they use. Whether this reputation is valid for all is open to doubt but most people in the property industry would recognise some truth in the statement. It is paradoxical then that, when the history of the valuation of investment property is reviewed, it is apparent that valuation techniques have constantly changed and adapted as the property market itself has evolved.

Consider market conditions at the turn of the last century. In 1900 leases of commercial property were very long; leases of 99 and 125 years were very common. Inflation was either unknown or not recognised, hence there were no mechanisms for adjusting the rent paid during the course of a lease. Property as an investment was very secure and stable, being similar to long-term government-issued fixed-interest securities (gilts). Valuation of property investments was accordingly simple, involving the capitalisation of the rent passing under the lease using an appropriate multiplier. This multiplier, the 'years' purchase' was simply the inverse of the initial yield, or annual return, on the property. This was easily observable from market transactions on property investments and directly comparable with the yield on gilts. As property was more illiquid, involved more management and was slightly riskier than gilts, it was obvious that property yields should be slightly higher to compensate the investor.

Let us now move forward 75 years and examine market conditions and valuation methods in 1975. Market conditions were obviously very different. Postwar inflation and changing tenant requirements had seen prime commercial leases settle into the pattern that would dominate the market for at least the next 15 years. Leases on this class of property were normally of 25 years' duration, with upward only rent reviews to the open market rental value, with the tenant being responsible, either directly or through a service charge, for all the running costs and maintenance on the property.

1

Valuation techniques had changed accordingly. The relationship between gilts and property was broken – property investments were now a growth stock something like equities, albeit with a peculiar income pattern. The straight initial yield approach was no longer reliable, particularly mid-way between rent reviews. The techniques were adapted to cope and it was in this period that term and reversion, layer and hardcore and equivalent yield methods were developed. New approaches were also developed, led by academics, although these were still essentially adaptations of the traditional techniques. UK academics such as Baum, Crosby and Mackmin promoted techniques such as the 'short-cut' discounted cash flow (DCF) and 'real value' approaches, both of which had the advantage of stripping out the inflationary element in the valuation and re-establishing the link between property and the overall return on fixed-interest securities.

This brief, superficial, review of property investment and valuation in 1900 and 1975 reveals that change in valuation does take place and that the drivers of change are primarily changes in market conditions. It also reveals some of the mechanisms of change, the two main mechanisms being the evolution and adaptation of existing techniques mainly by practising valuers, and the innovation and adaptation of techniques used in other disciplines mainly by academics. In the latter case, where the most novel valuation techniques develop, there has in the past been a gradual filtering and integration of the most useful methods into practice. This has come about in three main ways: firstly by academics producing articles for general consumption in the trade press, secondly by professional seminars and training, and finally by the natural integration of students, exposed to these techniques, into practice over time. The importance of this will be discussed further below.

Whatever the mechanism of dissemination of innovation, it must be re-emphasised that it is the market that is the driving force behind change. If the changes from 1900 to 1975 were striking it can be argued that the changes from 1975 to the present have been even greater and that the changes have been greatest in the last decade of the century.

The changes can usefully be viewed in terms of three categories, though there is some overlap and interrelationship between them. These categories are the property market environment, the economic environment and the technological environment.

Changes in the property market environment

The investment property market has seen some profound changes:

- Change in the security of income flow in property investments:
  - reduction in lease lengths during the 1990s (see Crosby and Murdoch, 1998);
  - wider use of break clauses and options;
  - volatility of firms in the occupier market (see below).
- Increasing complexity and volatility of property investments:
  - much of this is due to the factors identified above which have led to greater risk and uncertainty;
  - in addition, the use of lease incentives to secure tenants has made what was a relatively simple task, the establishment of an open market rental value (OMRV), extremely difficult.
- Increasing sophistication of property investors:
  - clients of valuations require more information from the valuer;
  - they are often willing to pay less for this additional information (see Havard, 1999a).

The results of these changes in the conditions in the market is that the market is more volatile, sometimes displaying more risk, but in particular it is more heterogeneous. Traditional valuation techniques work best in a market environment that is homogeneous, where there is more chance of finding good comparable evidence. In heterogeneous markets these chances are greatly reduced. This means that not only are the characteristics of the investments to be valued more complex, there is also less hard market evidence available to the valuer to support their valuation. In these conditions there is pressure to move to techniques that are less reliant on market evidence.

Changes in the economic environment

Equally importantly, the economic context in which property investments exist has also changed markedly:

- *Change to a low-inflation environment.* The sea change in macro-economic management in the western world towards inflation control priority appears to be long term. On the one hand this makes the job of the valuer easier, taking property back towards being a fixed-interest security. On the other hand, valuers are now more exposed: as we have seen, property income flows are less secure, less predictable. In the past, inflation has helped both the investor and the valuer, hiding mistakes made by both. That comfort is not likely to exist in the future.

- *Increasing covenant uncertainty.* Although certain things in business in the twentieth century have been remarkably durable, such as certain brand names (eg Coca-Cola), other aspects of business have shown greater volatility. Great business operations have risen to meet the needs of consumers yet fallen without a trace when they have failed to adapt to changing conditions (eg PanAm). This volatility seems likely to increase; twenty years ago one of the largest businesses in the world, Microsoft, essentially did not exist. In the UK, the Internet service provider Freeserve was launched in the middle of 1998. By the turn of the new century its market capitalisation was larger than UK giants such as British Airways. The impact of things such as e-commerce are only starting to be felt. The point is that the certainty of 'blue chip' covenants in property so important to investors and valuers may be illusory. The degree of uncertainty and volatility in property may be greater than has been anticipated.

    Similarly, the requirements of these businesses in terms of the space they require is likely to change as quickly. The impact of changing air conditioning, telecoms and IT requirements made many office structures obsolete in the last twenty years – the introduction of e-commerce and virtual companies is likely to add to the pace of obsolescence. All of this has consequential effects on the task of valuers and appraisers, given what their role is.

Changes in the technological environment

Finally the technological environment in which valuations are conducted has also changed. Taking a narrow view of technological change (ie that the changes in the business environment, outlined above, are mainly due to changes in technology), there are two principle effects of the IT revolution:

- *The amount of information available to players in the property market.* This, like the inflation factor, is a double-edged sword: on the one hand there are greater amounts of material to analyse. This encourages the use of new methods and levels of analysis. We have seen this in the rise of the Investment Property Databank (IPD), the use of IPD data by investors and analysts, and the growth and development of property market forecasting. On the other hand, investors are aware of these changes and themselves require more sophisticated approaches.

- *The change in accessibility of the technology available to valuers to analyse data and undertake quite complex analysis.* The advent of cheap, powerful desktop PCs with the software to run them has greatly extended the ability of valuers.

Summary

The changes in these three environments provide great incentive and impetus for valuation techniques to change, just as they have done in the past. In the past, however, the pace of change has been relatively slow – certainly compared with what has happened in the last twenty years the pace was positively pedestrian. However, the pace of change mirrors most things in human society, creating problems for both the evolution of old methods and the development of new techniques. Normal methods of dissemination have struggled to keep up. At the same time many leading academics have been forced to neglect mainstream valuation thought as research money and academic kudos has led to a concentration of their efforts in the fields of market analysis and forecasting. Development in investment valuation thought has occurred but its dissemination has been rather limited.

This then is the rather bold intention of this short book. It attempts to review where we are in valuation practice and theory, not in great depth but in an easy to understand, practical way. It aims to bridge some of the gap between academic and practical valuation by reviewing some the principal traditional, contemporary and 'advanced' valuation techniques and methods available to the investment valuer at the dawn of the twenty-first century.

PART 1

The Valuation Toolkit

A Review of the Traditional, Contemporary and Potential Capital Valuation Techniques

The intention of this book is that it should allow the reader:

- to obtain at least an overview of the key techniques and approaches available in their valuation 'toolkit';
- to understand the strengths and weaknesses of each;
- to appreciate when and where the various techniques are best applied.

This is important as it should be stressed that the core techniques which have been used in valuation for generations should not be discarded under current market conditions. The principles of valuation and the methodologies used in 1900 can be successfully applied in the year 2000 and beyond. The problems arise where traditional techniques are used without thought, or where they are used in complex circumstances which they were not designed to cope with. In these cases there is a greater risk of the valuation failing.

There is therefore a need to look at both familiar techniques and methods and also some of the newer, more advanced methodologies because each has their place. This means that 'advanced' does not always means 'superior'. The most complex, technologically advanced method might not be the best way of tackling a simple problem just as the simple way almost certainly would not be the best for tackling complexity. The story about the space pen is brought to mind. This story may be an urban myth but the meaning is appropriate. Early in the manned space programme, NASA scientists realised that astronauts were going to have problems writing notes in zero gravity as conventional pens were gravity fed. After many years of expensive research and development, NASA scientists came up with the space pen that had a complex integral pump activated by the pressure of the point. It worked absolutely beautifully. Russian space scientists of course had the same problem. They solved it in five minutes at near zero cost. How? They issued their cosmonauts with pencils! The moral of the story is clear: don't discard the simple unless you have to.

There is a clear division between academics and practitioners in this area. Academics find many of the traditional methods of valuation to be logically unsound and mathematically incorrect. Practitioners often feel that academic developments are unnecessarily complex and too theoretical to be of practical use in the marketplace. The truth lies somewhere between the two views. Academics perhaps underestimate both the difficulty of the environment in which practising valuers work and also the degree of skill, judgement and self-reliance that valuers possess. Practitioners often do not appreciate the flaws of the techniques they use nor do they have a real understanding of alternative techniques that would allow them to be confident in their application in practice.

This part of the book outlines the various methods and approaches available to the valuer, starting with the simplest, traditional methods of valuation then moving onto the more complex. Inevitably, this involves moving from what are clearly identifiable as 'valuation' techniques to those that are largely viewed as being 'investment appraisal' methods. In the past it has not been felt appropriate to apply investment appraisal approaches to valuation because each investment will be viewed differently by each individual investor. It was thought that if valuers were asked to use such methods, the range of different alternative constructional assumptions would lead to different answers being produced by different valuers. In fact, whether a group of valuers using appraisal techniques would show a wider variance in the valuations they produced than a similar group using traditional valuation techniques has not been tested. Valuers using traditional methods have illustrated a wide degree of variance (Huchinson *et al*, 1996; Havard, 1999a) in the results of their valuations. It is hard to believe that the use of appraisal techniques would lead to a higher degree of variability. Whatever, there are developments in the marketplace that make it more difficult to use traditional techniques based on comparable evidence all the time. As property investments become more complex and diverse, as future cashflows become more uncertain, then there will be an increasing need to use techniques that look ahead or are not tied purely to comparisons. The use of 'investment appraisal' techniques for valuation in the future is inevitable.

Chapter 3

Traditional Valuation Techniques

Definitions and principles

All of the investment valuation methods are based upon the principle of discounting:

> **2.1.1** The underlying principle of assessing the price or worth of an investment is to discount the net benefits and liabilities expected to be produced by that investment over its lifetime at a specified rate of return or discount rate.

> **2.1.2** Discounting is the process of finding the present value of the right to receive the expected future income flow. The cash flow may be in the form of a regular income or a future capital receipt, or a combination of the two.

RICS (1997), p. 11

The valuation profession uses a variety of different models to value investment property in the UK. Some of the differences in approach are due to the specific circumstances of the property being valued, particularly the lease and income flow structure, but there is also a difference due to the treatment of future growth potential, ie whether it is implicitly or explicitly allowed for.

Once again a direct quote from the RICS information paper illustrates the issue:

> **3.1.2** In a situation where rental income and capital value are expected to increase over time, investors in property are willing to accept initial yields on their investments below the yield they could achieve from a risk-free, no-growth investment such as government stock. Effectively they accept a lower yield initially, in expectation of higher cash flows and/or an increase in capital value in the future.

> **3.1.3** In using the traditional implicit capitalisation model, the valuer is deriving the appropriate all-risks yield from market evidence of other property transactions. The all-risks yield is a convenient measure for the analysis and valuation of similar rack-rented investments. Effectively, the all-risks yield states that such investments customarily sell for a certain multiplier of the rental income. Adjustments to the all-

risks yield to reflect differences between comparables and the subject
property are made subjectively.

RICS (1997), p. 12

It goes on to give a further clarification of what the all-risks yield
(ARY) actually is:

> **3.2.1** The all-risks yield is an (annual) IRR [internal rate of return] of a
> non-growth cashflow. Rental growth, obsolescence and the resale value
> are reflected implicitly.

RICS (1997), p. 12

Traditional techniques, therefore, are discounting techniques, albeit
with all assumptions not made explicitly in the cash flow but
implicitly in the yield used.

How the ARY is used in practice depends upon the income
pattern of the property investment being valued. The original
valuation model, as we have seen, evolved at a time when there was
no need to consider inflation. In situations where the property
would let at a higher rent than the rent contracted under the lease
and there is a mechanism for adjusting the lease rent to a higher
figure at some point in the future, then this clearly should be
reflected in the price an investor would be willing to pay. A result of
this is that the initial, or income, return on the property is reduced.
While this is perfectly normal in investment terms this does create
some problems in valuation which will be examined below. The
result of this is that the initial ARY approach to valuation of these
types of property, where there is 'reversionary' value, is undertaken
with a modification to the initial yield approach. We will review the
strengths and weaknesses of these different approaches in the next
few sections, starting with the initial yield approach.

Initial yield

Most of the pricing of investment assets in the UK is still done
using this traditional approach. Its simplest form is found with
'rack-rented' property, where the rent passing on the property is
equivalent to the rent which would be achieved if the premises
were vacant and to let on the open market on terms that the market
would find 'normal'. Here the equation for valuation or pricing the
asset is simply:

$$V^0 = R / i$$

where:

V^0	=	value at date of valuation
R	=	rental value or income
i	=	all-risks yield

or alternatively:

$$V^0 = R \times 1/\ i$$

where the formula $1/\ i$ is that which derives a rental multiplier or years' purchase (YP).

Traditionally, the initial yield is market derived, as we have already observed, by the analysis of transactions of similar property investments that have occurred in the open market.

Even with this most simple approach problems can arise with valuations under certain conditions. Assuming that the rental value can be established to be equivalent to the rent passing, the model requires that comparable sales of similar investment properties be examined to arrive at an appropriate yield figure. In reality it is relatively rare that a direct comparable can be found even in good markets. As the RICS information paper states:

> **3.2.2** The all-risks yield is adjusted to reflect differences between comparables and the subject property. It has to reflect a multitude of factors such as security of income, ease and cost of selling, ability to use as loan collateral, management costs, depreciation and rental growth.

If a sample of properties within any city centre is examined and the diversity of the quality of tenants, buildings and leases is recognised, the complexity of even the simplest of valuations can be appreciated. This is recognised in the RICS document:

> **3.2.3** One problem with the all-risks yield is the difficulty of assessing transactional evidence and rationalising the adjustment required to apply it to other investments. Even a minor difference such as a rent review for one property being closer than for another may make comparison difficult.

> **3.2.4** Where there are sufficient sales transactions on similar properties in the market, it is possible to build up a picture of market sentiment to be reflected in the choice of an appropriate all-risks yield for the subject property. However, where there is limited comparable evidence for the specific cash flow pattern to be valued, the absence of rational analysis in the all-risks yield approach makes it difficult for clients to understand.

RICS (1997), p. 12

In theory, the initial yield approach can be used to value all income-producing property. In practice, as the complexity of the property increases, then the chances of finding sufficient comparable evidence decrease and the risks of using the approach increase.

Initial yield valuation is simple and problem-free as long as certain conditions exist:

- that the subject property is let at the current OMRV;
- that there are a substantial number of transactions on similar properties in the open market;
- that the future income flow of the property is relatively secure and simple.

If these conditions are met then the simple, initial yield approach has much to commend it, principally in that the value of the property investment can be directly related to transactions of similar properties with little adjustment of the evidence. It does have the inherent problem of all of the traditional valuation techniques in inflationary and deflationary environments, namely that the yield on the property does not reveal the expected total return on the property. It also does not reveal the assumptions made by the valuer about the net rate of growth expected in the future.

Initial yield valuations will have problems where:

- there are insufficient transactions in the marketplace;
- the property is not let at the current OMRV;
- the investment being valued has characteristics that are different from other properties in the marketplace. These differences can be physical, or may be differences in the quality of the covenant of the tenant, or relate to the lease terms under which the property is let.

Where these conditions exist there are inherent dangers in using this approach. For example, where the OMRV differs markedly from the lease rent, the initial yield will fall to such an extent that minor variations in yield make a major difference in value. This can be illustrated by an example. A property has been sold in the open market for £690,000 (excluding costs). The lease rent is £20,000 pa, fixed for the next three years. The current open market rental value is £50,000 pa. This produces a low initial yield of just under 2.9%. Let us say we are attempting to value an otherwise similar property, also let at £20,000 fixed for three years, but where the OMRV is £40,000 pa. The transaction is our only piece of evidence. We probably need to adjust the evidence to suit our subject property, but which way? There is more value in the reversion of the comparable but there is relatively less risk in the subject. We might put a yield just below the comparable at 2.75% or at the same level or perhaps just above at, say, 3.25% with equal justification. The three resulting valuations are illustrated in the box below. As can be observed, there is over £100,000 difference in the highest and lowest valuations in a property worth less than £700,000.

This extreme sensitivity to yield choice led to the development of modifications to the traditional valuation model. These will be reviewed below.

Valuing reversionary and other uneven investment types

The normal UK lease structure for good quality commercial property is for property to be let on relatively long leases with periodic rent reviews to adjust the rent passing on the subject premises to that which would be achieved in the open market. (At least this is the intention of review clauses. In practice open market lettings and rents achieved at review for properties with otherwise similar characteristics can differ quite markedly.) The normal length of lease has shortened markedly since the late 1980s: 10–15 years seems to be the norm for the best office properties with the traditional 25-year lease period being mainly confined to the best quality retail premises. Rent reviews in the UK market for commercial properties are invariably for periods in excess of one year and indeed five years is the norm.

This lease structure means that the rents on commercial property are prone to becoming out of step with rents achievable on similar properties let in the open market. Over most of the period from the early 1960s onwards (when rent reviews commenced) to the present, market rents (as opposed to lease rents) with inflationary and real growth in values (ie growth due to excess demand over supply) tended to quickly outstrip rents set only a short period before. The resulting investment is termed 'reversionary', that is the income received is below that which is achievable in the open market. An investor in the market for such an investment would make their bid with this situation in mind. A number of different valuation models have been developed to deal with this type of investment. These will be reviewed below.

Reversionary investments under UK commercial leases have two components of value. The first arises from the contracted rent, the

Valuation 3.1 Initial yield @ 2.75%

Lease rent	£20,000	
YP perp. @ 2.75%	36.36363636	£727,272.73
Less costs[1]	5%	£34,632.03
		£692,640.69

1   Generally valuers in practice make an allowance for purchasers' costs in calculating the valuation of a property. In order to achieve the stated yield on a property such a deduction will be made. The costs include stamp duty, solicitors' and agents' fees. The deduction can be calculated by taking a percentage of the gross valuation though strictly this will overstate these costs. In this book, where allowed for, the costs will be calculated by dividing the gross valuation by 1 + Purchasers' cost percentage, ie 1 + 5% in this case.

Valuation 3.2 Initial yield @ 2.90%

Lease rent	£20,000	
YP perp. @ 2.90%	34.48275862	£689,655.17
Less costs	5%	£32,840.72
		£656,814.45

Valuation 3.3 Initial yield @ 3.25%

Lease rent	£20,000	
YP perp. @ 3.25%	30.76923077	£615,384.62
Less costs	5%	£29,304.0
		£586,080.5

lease rent. This income flow is guaranteed, at least as long as the lease continues. The security of this income flow is usually reinforced by 'upward only' or ratchet clauses in the rent review clause. These clauses do not allow the rent to fall even if the OMRV is below the rent passing under the lease. The second component of

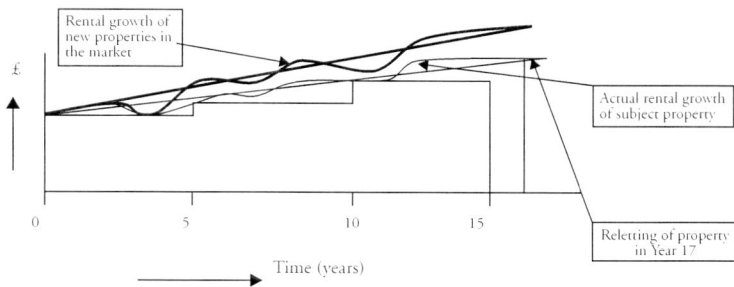

Figure 3.1 Income flow pattern from a 15-year FRI lease with rent reviews to market every five years. (Straight lines represent long-term trend lines for rent.)

value is the unsecured part of the income, reflecting the difference between the passing rent and the OMRV.

In a 15-year full repairing and insuring (FRI) lease with rent reviews to market every five years the income flow pattern of the investment will be as illustrated in Figure 3.1.

This stepped type of income flow is typical of a UK commercial property investment. At year zero, the actual pattern of the income flow will be unknown. In particular, the actual change in the rent achieved at rent review and what will actually happen at the end of the lease is uncertain. How these uncertain factors are allowed for distinguish the valuation methods used to value these properties. In the traditional approaches, the assumption about the future income flows are made implicitly in the yield used. In the contemporary and advanced approaches, some explicit assumptions are made about these factors.

Figure 3.1 illustrates a possible income pattern from the start of an investment. In between the start of the lease and the first rent review and in between the subsequent rent reviews, the rent passing under the lease may not equal the lease rent. Until the early 1990s this normally meant that the passing rent was below the OMRV. How this situation is handled in valuation models distinguishes the various approaches.

Throughout the next few sections we will be considering how we might value the following property:

> A modern office property let to a single tenant two years ago on a 15-year
> FRI lease with rent reviews every five years to the current OMRV. The
> contracted rent is £100,000 pa. The current OMRV is £110,000 pa.
>   Similar investments, albeit let at the current OMRV, have sold
> reflecting yields of 7%.

Our review of the valuation models will start with the traditional
growth-implicit models. There are a number of different possible
approaches but they break down into two basic families, one of
which treats the investment as two vertical slices, the other as two
horizontal slices. These are, respectively, the term and reversion
and the layer and hardcore methods of valuation. These
approaches to valuation are illustrated in Figure 3.2.

Two features of the models should be noted. First, only two
periods are considered. After the next rent review the property is
treated as if it were a rack-rented freehold, ie it is valued assuming
the cash flow will continue into perpetuity. Second, there is no
projection of rents: all rental values are taken as existing at the time
of the valuation.

The relative merits of the two basic models will be considered in
the next section.

Term and reversion equivalent yield

There are a number of variants of both the layer method and the
term and reversion approach. Principally the variation depends
upon whether the same yield is used for both parts of the investment or not. Where the same yield is used throughout this is
termed an 'equivalent yield' approach. Where different yields are
used, these are referred to as 'split yield' approaches.

An equivalent yield approach to the valuation of our base
property is illustrated in Valuation 3.4.

The equivalent yield approach essentially uses a growth implicit
internal rate of return (IRR) to calculate the net present value of the
investment. It should be noted that no projection of income
expected at rent review in three years' time has taken place; the
£110,000 is the current OMRV. Any growth in income is allowed for
in the yield used.

The yield has been adjusted from 7% in the same way that a
valuer in practice might. This adjustment tends to be done

Term and reversion

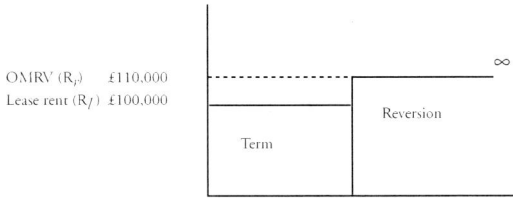

OMRV (R_r) £110,000
Lease rent (R_l) £100,000

Reversion

Term

∞

Formula: $\quad R_l \times \dfrac{\dfrac{1}{1-(1+i)n}}{i} + R_r \times \dfrac{1}{i} \times \dfrac{1}{(1+i)n}$

Layer and hardcore

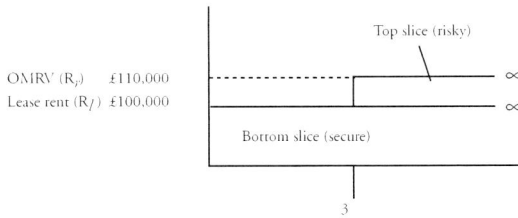

Top slice (risky)

OMRV (R_r) £110,000
Lease rent (R_l) £100,000

Bottom slice (secure)

∞

∞

3

Formula: $\quad R_l \times \dfrac{1}{i} + (R_r - R_l) \times \dfrac{1}{i} \times \dfrac{1}{(1+i)n}$

where:

i =   capitalisation yield
R_l =   lease rent
R_r =   current open market rental value
n =   number of years to the rent review.

Figure 3.2 Comparison of term and reversion and layer/hardcore techniques of valuation

subjectively, taking the rack-rented yield as a benchmark and reducing or increasing the yield from the benchmark figure according to perceived market sentiment or the characteristics of the investment. It should be noted that the same comments made about

---

Valuation 3.4 Term and reversion equivalent yield

Assumptions/facts

Rent	£100,000		
OMRV	£110,000		
Reversion	3		
RR yield	7%		
Equivalent yield	6.50%		

Calculation

Term		£100,000	
YP 3 yrs @ 6.50%		2.648475511	£264,848
Reversion		£110,000	
YP perp. @ 6.50%	15.38461538		
Def. 3 yrs @			
6.50%	0.827849092	12.73613987	£1,400,975
			£1,665,823

Valuation 3.5 Split yield term and reversion

Assumptions/facts

Rent	£100,000		
OMRV	£110,000		
Reversion	3		
RR yield	7%		
Initial yield	6.00%		

Calculation

Term		£100,000	
YP 3 yrs @ 6.00%		2.673	£267,301
Reversion		£110,000	
YP perp. @ 7.00%	14.28571429		
Def. 3 yrs @ 6.00%	0.839619283	11.995	£1,319,402
			£1,586,703

---

the problems of adjusting yields from comparables with the ARY also apply here, with the added complication of the variation in investment pattern which can arise from different periods to reversion and different amounts of reversionary value. As Baum, Mackmin and Nunnington (1997) note: 'Valuers are still faced with the problem of finding comparables. The more unusual the patterns of income the more difficult it is to judge the correct capitalisation rate.' This is made increasingly difficult by the fact that investments with variable income flows are relatively difficult to analyse to derive equivalent yields without resort to computers.

Split yield term and reversion

A split yield term and reversion approach to valuing the subject property is presented as Valuation 3.5.

This is the alternative way that some valuers approach this kind of problem. The valuer here has chosen to manipulate the term yield, though some valuers will manipulate both term and reversionary yields. The reason that this is done is 'to reflect some personal view on security of income' (Baum, Mackmin and Nunnington, 1997, p. 64). Baum *et al* (1997) go on to give the historical background to this:

> In the 1930s the depression made any contracted rent a better or more secure investment than an unlet property and there was real fear that at the end of a lease the tenant could seek to redress the position by requesting a lower rent or by vacating. Logic suggested the use of a lower capitalisation rate for secured contracted rents than secured reversions.

They note that this situation continued until the 1960s when the effect of inflation and the inclusion of rent reviews led to property being perceived to be less comparable with gilts but more like equities in the growth potential of capital value and income. Due to this a reverse yield gap developed. In the 1930s property yields were related to gilt yields but had a premium attached to them to reflect the increased risk, management problems and lack of liquidity as an asset. From the 1960s the yields on property dipped below those on gilts. This was because of the inherent growth potential in the asset. Despite this, the methods used to value properties did not change and the tendency to use more than one yield in a valuation persisted for reversionary properties, the argument being that the term yield should be lower as the tenant paying a rent below the market rent was less likely to default.

There are inherent dangers with this technique. Firstly the adjustment tends to be done subjectively as here, and some

---

Valuation 3.6 Equivalent yield layer or hardcore approach

Assumptions/facts

Rent	£100,000
OMRV	£110,000
Reversion	3
RR yield	7%
Equivalent yield	6.50%

Calculation

Bottom slice		£100,000	
YP perp. @ 6.50%		15.38461538	£1,538,462
Top slice		£10,000	
YP perp. @ 6.50%	15.38461538		
Def. 3 yrs @ 6.50%	0.827849092	12.73613987	£127,361
			£1,665,823

Valuation 3.7 Split yield layer or hardcore approach

Assumptions/facts

Rent	£100,000
OMRV	£110,000
Reversion	3
RR yield	7%
Bottom slice yield	6.00%
Top slice yield	8.00%

Calculation

Bottom slice		£100,000	
YP perp. @ 6.00%		16.66666667	£1,666,667
Top slice		£10,000	
YP perp. @ 8.00%	12.5		
Def. 3 yrs @ 6.00%	0.839619283	10.49524104	£104,952
			£1,771,619

---

variation from the benchmark equivalent yield valuation can be
noted, although this tends to be relatively minor with the split yield
term and reversion approach where the term is relatively short as

this is only a marginal part of the total value (here the difference is 5%). Some valuers choose to defer the reversionary portion of value at the reversionary rate, which produces a slightly different value and is also technically incorrect. In addition, as Bowcock (1983) notes, variable rates can produce some illogical results in a valuation, illustrating this with a valuation of two properties both let at £100 and both with OMRVs of £105 but one with five years to the reversion and one with ten years. The property with a ten-year reversion produces a higher value when yields of 9% are applied to the terms and 10% to the reversion for both when logically the reverse should be true.

What this illustrates is that the traditional methods are mathematically flawed – they are not doing what the valuer believes they are doing. This comment applies to all of the traditional methods. This does not mean that they should not be used for valuation – indeed, while the property sector as a whole is using the same methods to assess the value for property it is likely that these methods will be best at predicting the exchange price of a property investment. Valuers should be aware of these flaws, however, and be cautious about the use of these approaches.

Equivalent yield layer or hardcore method

This valuation could have been set out using a layer or hardcore model (see Valuation 3.6) with the same outcome. Using this method, a horizontal split of income is assumed, the lower layer representing the 'secure' part of the income, ie the rent which the tenant has contracted to pay under the lease. The upper layer is the unsecured part of the income, that which the valuer expects to be achieved at the next rent review or lease renewal.

The key advantage to this for valuers is that the mathematics are slightly simplified with both top and bottom slices of the income using a YP in perpetuity to capitalise the income flows, with the latter being deferred at an appropriate rate.

Split yield layer or hardcore method

A split yield layer or hardcore approach to the valuation of the subject is presented as Valuation 3.7.

The same comments made about the split yield term and reversion apply to this model although variations in the yield of either the top or bottom slice are even greater. Relatively minor

differences in the bottom yield in particular can produce a very
different outcome in the valuation. This method is thus very
vulnerable to yield selection and is not recommended.

Summary of traditional approaches

The traditional methods of valuation have both strengths and
weaknesses. The strengths relate to familiarity and ease of use.
Some of the problems are technical, as identified in the text above.
Many of these can be avoided by adopting the equivalent yield
approach. There are still mathematical problems that reflect the fact
that these methods are compromises or adaptations of the basic
initial yield approach to meet circumstances that the original
technique was not designed to deal with. There are also
mathematical inconsistencies; for example, the term and reversion
approach overvalues the term and undervalues the reversion. All
the methods have the basic problems of the traditional approaches
in that they do not reveal the assumptions made about growth or
total return to the recipient. The methods also still rely on there
being good comparisons available in the investment market,
though they are far less sensitive to yield choice than with the
initial yield approach. Where the traditional approaches really
suffer is where the income flow becomes complex or uncertain or
both. The traditional methods were really exposed by over-rented
property investments. These will be considered in detail in Part 2.

Chapter 4

Contemporary Income Valuation Techniques

'Short-cut' DCF approach

One of the main criticisms of the conventional valuation models is that they do not reveal the underlying assumptions that make up the final valuation. They do not illustrate the total rate of return that is being assumed an investor would seek nor the required growth rate that is necessary to achieve this figure. The 'short-cut' DCF model overcomes this problem and has some technical advantages in dealing with both reversionary and over-rented properties.

The model has been criticised on the grounds that the discount rate or equated yield is derived subjectively, there being no market evidence as to what investors' requirements regarding this component are. However, the choice of this rate makes hardly any difference to the final value. This is because the growth rate used is only implied through looking at the relationship between the ARY (usually referred to as k), the equated yield e and the rent review period t. The growth rate can alter with variations in either k or e but as the discount rate also changes with the latter, it is only the former that will affect the final value. As long as the ARY for rack-rented properties can be reasonably securely established the method is thus relatively insensitive to yield choice. The problems of establishing the ARY mentioned above do, however, still apply.

The implied annual growth rate can be calculated using a number of formulae, one of the most popular being:

$$(1 + g)^t = \frac{\text{YP in perp. at } k - \text{YP}_t \text{ years at } e}{\text{YP in perp. at } k \times \text{PV}_t \text{ years at } e}$$

where:

k = the capitalisation rate expressed as a decimal (ie the ARY)
e = the overall return or equated yield expressed as a decimal
g = the implied annual growth rate
t = the period between rent reviews in years

e is normally assumed to be a percentage above the rate on government issued long-term securities (gilts). This will be discussed

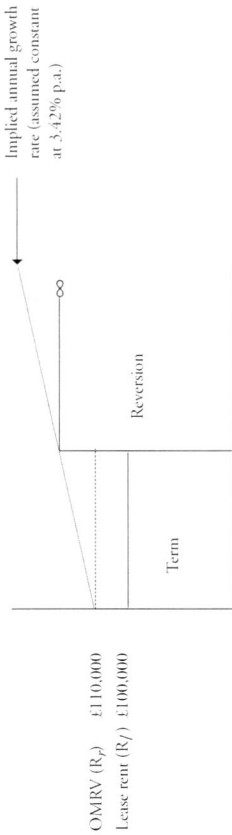

OMRV (R_r) £110,000
Lease rent (R_l) £100,000

Implied annual growth rate (assumed constant at 3.42% p.a.)

Term   Reversion   ∞

Formula: $R_l \times \dfrac{1-(1+e)^{-t}}{e} + [R_r \times (1+g)^t] \times \dfrac{1}{k} \times \dfrac{1}{(1+k)^t}$

where:

k = capitalisation yield (i.e. the ARY)
e = equated yield
g = implied annual growth rate
R_l = lease rent
R_r = current open market rental value
t = number of years to the rent review.

Figure 4.1 Assumptions contained within the 'short-cut' DCF valuation

further below. In this case we will assume that e is 10%.

Using the details from our base property, this gives the following values in the formula:

$(1 + g)^5$ = YP perp. at 7% − YP 5 years at 10% / YP perp. at 7% × PV 5 years at 10%

           = 14.286 − 3.791 / 14.286 × 0.621

           = 10.495 / 8.872

           = 1.182987612

Therefore g = 3.4209%   (ie $(1.182987612)^{(1/5)} - 1$)

It should be stressed that the implied growth rate g is not a forecast but an indication of an expectation of growth that must be achieved to meet the performance parameters set. The assumptions contained within the valuation are presented in Figure 4.1.

As can be observed, the method is more explicit about the basic assumptions contained within the valuation, at least until the first rent review. The rent capitalised at this point is the OMRV inflated at the assumed long-term implied growth rate. Other than this variation, from this point the valuation is the same as with the traditional ARY approaches.

The valuation of our base property is illustrated in Valuation 4.1. The outcome of this valuation is discussed in the next section on 'full' DCF approaches. It is important that both of these sections are read together to get a full appreciation of the techniques involved.

'Full' discounted cash flow approach

This approach is still infrequently used in practice in the UK for valuation although it is the primary tool of investment analysis. However, there is anecdotal evidence that the use of 'full' DCF approaches is on the increase, in itself an indication of changing market conditions and requirements. The 'full' DCF approach is widely used in the US and has been for many years, largely because market conditions in the US encouraged its development. The US commercial market is characterised by short leases, non-recoverable outgoings for maintenance, void periods, periods of partial occupancy, multiple occupancy, etc. What it amounts to is complexity, diversity and uncertainty, conditions that favour DCF over the traditional methods of income valuation. It should be noted, however, that US appraisers rarely, if ever, use DCF alone in their valuations. In addition, DCF is obviously not flawless. It can be manipulated and misleading, and did not prevent such crises as

---

Valuation 4.1 'Short-cut'DCF approach

Assumptions/facts

Rent	£100,000		
OMRV	£110,000		
RR yield	7%		
Reversion	3		
Equated yield	10.00%		
Implied annual growth rate	3.4209%		

Calculation

Term		£100,000	
YP 3 yrs @ 10.00%		2.486851991	£248,685
Reversion		£121,679	
YP perp. @ 7.00%	14.28571429		
Def. 3 yrs @ 10.00%	0.751314801	10.73306858	£1,305,994
			£1,554,679

---

the 'savings and loan' debacle or predict the crash in property prices of the early 1990s that affected all global markets.

What is the difference between 'full' DCF and the 'short-cut' approach described in the previous section? The answer is that it differs from the latter – and indeed from all the other methods described – in its lack of reliance on market evidence for yields and the rental forecasts used in the valuation. Neither the discount rate, the forecast of growth rate or the assumption regarding terminal yield are drawn directly from market comparisons, though they are usually based on an analysis of the market. It is this characteristic which perhaps has restricted the adoption of 'full' DCFs into the UK market.

Let us take the treatment of growth and terminal yield. In the 'short-cut' DCF model, the growth rate used is implied, ie it is derived from the implied assumptions of the participants in the market arising from market transactions. In our base example the rack-rented ARY was 7% and investors were assumed to be seeking a 10% annual return. If the rents on the property investments were adjusted annually then the required growth rate, ie the one that investors must be expecting, is 3% pa (the difference between the income return of 7% and the total required return of 10%. Because the property had a five-year rent review cycle, the required rate of growth has to be higher to compensate the investor for the delay. In this case the required growth rate is 3.4209%.

This implied growth rate is *not* a forecast growth rate. Neither is it an *actual* rental growth rate. It may be that actual rental growth on similar but new properties in the market may be, say, 5% over the same period. What it is, is the implied long-term average income growth rate for the investment. It includes an allowance for the deterioration of the competitiveness of the investment over time. We know that certain types of investment depreciate over time, mainly due to the wear and tear of the building and changes in fashion or user requirements (obsolescence). This is particularly true of offices and industrial buildings. In the actual property market this is reflected in two ways over time: a reduction in the rental growth rate relative to new buildings and an increase in yields over time, reflecting investors' views of the future.

Both of these factors are allowed for in the 'short-cut' DCF approach. Indeed, it is very important not to manipulate the yield on the reversion to reflect a fall-off in competitiveness or any changes in market conditions expected where the implied growth rate is used. The resulting valuation will be mathematically inconsistent, will not produce the same answer as that produced by a traditional approach and will also probably be wrong!

The 'short-cut' DCF approach is essentially a traditional valuation with its internal assumptions made explicit. In contrast, the 'full' DCF approach usually requires the valuer to forecast the actual likely cash flows per period over a period into the future. This may require the actual rental growth (or change in rental values – the market may fall). Similarly, the actual sale yield of the investment, taking into account the relative depreciation of the property, may explicitly be required. Each individual cash flow is then discounted back to present-day values using an appropriate discount rate.

The reader will probably have noted the use of the words 'may' and 'usually' in the previous paragraph. The reason for this is that there is more than one type of DCF model – in fact all income valuation models, traditional or otherwise, are discounted cash flows. How each type varies depends on how explicit the assumptions about the future are. In the traditional approach, assumptions about growth and depreciation are contained within the ARY. In the 'short-cut' approach, some of the implicit assumptions are made explicit. Beyond that, the full approach

Valuation 4.2 'Full'DCF approach to valuation of subject

Assumptions/facts

Rent	£100,000
OMRV	£110,000
RR yield	7%
Reversion	3
Discount rate	10.00%
Forecast annual growth rate	5.0000%
Terminal yield	9%

Calculation

Year	OMRV [1] £	Lease rent £	Sale value £	Cash flow £	Discount factor	DCF £
0	110,000					
1	115,500	100,000		100,000	0.9091	90,909
2	121,275	100,000		100,000	0.8264	82,645
3	127,339	100,000		100,000	0.7513	75,131
4	133,706	127,339		127,339	0.6830	86,974
5	140,391	127,339		127,339	0.6209	79,067
6	147,411	127,339		127,339	0.5645	71,879
7	154,781	127,339		127,339	0.5132	65,345
8	162,520	127,339	1,805,779	1,933,118	0.4665	901,814
					NPV	**£1,453,765**

1   Note that this column is only used to calculate the rent at review. It is not included in the cash flow calculation.

makes more things explicit, and how explicit depends upon how far the person constructing the model wishes to go. For example, rental growth might be represented as a long-term average rate over the valuation period, or else an attempt might be made to forecast the actual cycle of rental performance over the period. The DCF model can be made more and more sophisticated. The problem with this is that the greater the number of explicit assumptions that are made, the more likely it is that they will be different from those made by other valuers and thus that the valuation itself will differ from valuer to valuer.

The period of analysis can in theory be anything, but the most common cash flow periods used are monthly, quarterly or annually. Although the normal period of income receipts from commercial property in the UK is quarterly in advance, it is more common to use annual cash flows in the discounted cash flow models for commercial property in the UK. Discounted cash flow models work best where there is a known, finite time horizon. Freehold commercial investment property, in contrast, is perpetual; potentially such property can produce income for an infinite period of time. This is allowed for in DCF models used for property investment analysis and valuation work by assuming a notional sale of the property at an appropriate point in the cash flow.

In this case the property's income flows are examined in detail up to the second rent review in year 8, after which a sale is assumed at a yield of 9%, capitalising the year 9 income, which is assumed to be the OMRV as at the end of year 8 (see Valuation 4.2). In the initial model, an average growth rate of 5% is assumed over the valuation period.

This is one way of constructing the model. An alternative way is displayed in Valuation 4.3. Here the valuation assumptions are even more explicit. A forecast of the cycle of rental growth rates over the next eight years is included instead of the average rate used in the model in Valuation 4.2. Similarly, an assumption has been made about the relative rental growth performance of the property with regard to how the property will age: in the early years the property is assumed to grow almost as quickly as the new properties in the market, while as it ages, growth is not as fast.

The NPV is hardly different from the initial appraisal, but is different nonetheless, illustrating that the assumptions made are not exactly the same between the two approaches. It can be imagined that between valuers the differences might be more marked.

These then are the main characteristics of the DCF model, which reveal a lot about its advantages and disadvantages as a method of valuation. The advantages relate to its flexibility, its ability to deal with complexity and the explicit nature of the assumptions that can be made in its construction. Its disadvantages relate to its variability. The advantage of the 'short-cut' DCF is that it is derived from the analysis of market behaviour evident from market transactions. The analysis should be uniform across the market. With the 'full' DCF approach the assumptions may well vary from valuer to valuer. Despite this, there are clear attractions in using this approach in valuation. As long as the valuer is aware of these potential problems

Valuation 4.3 Alternative approach to 'full' DCF valuation of the subject

Assumptions/facts

Rent	£100,000
OMRV	£110,000
RR yield	7%
Reversion	3
Discount rate	10.00%
Forecast annual growth rate	See table below
Terminal yield	9%

Forecast annual growth rate

Year	Forecast growth in market rents (%)	Relative depreciation (%)	Effective rental growth for subject (%)
0			
1	3	97	2.91
2	5	97	4.85
3	6	95	5.70
4	8	95	7.60
5	4	92	3.68
6	6	92	5.52
7	5	90	4.50
8	5	90	4.50

Calculation

Year	OMRV £	Lease rent £	Sale value £	CF £	DF	DCF £
0	110,000					
1	113,201	100,000		100,000	0.9091	90,909
2	118,691	100,000		100,000	0.8264	82,645
3	125,457	100,000		100,000	0.7513	75,131
4	134,991	125,457		125,457	0.6830	85,689
5	139,959	125,457		125,457	0.6209	77,899
6	147,685	125,457		125,457	0.5645	70,817
7	154,331	125,457		125,457	0.5132	64,379
8	161,275	125,457	1,791,950	1,917,406	0.4665	894,484
					NPV	**£1,441,953**

and can make reasonable assumptions, there is no doubt that the technique can assist the valuer in many circumstances.

Much of the criticism of using the 'full' DCF technique is, as noted, based upon the requirement of the valuer to make judgements and assumptions about the future and that these may well vary from valuer to valuer. This has tended to result in both practising valuers and academics shying away from using 'full' DCFs in valuation. Academics in the UK, in particular, have shown a clear preference for working with those approaches which are the most mathematically accurate, defensible and comparable with the traditional techniques. This is understandable but seems to have been made on the assumption that the investment market is rational. In fact, even using traditional techniques, valuers are constantly making value decisions which often already differ from valuer to valuer. In commercial valuation, the evidence available to the valuer is usually inadequate and open to quite different interpretation and application. Evidence for this can be seen in the studies of valuation variance that raise doubts about what an acceptable range of difference is between valuers. In summary, what this amounts to is that, although the theoretical problems of variability with the 'full' DCF are accepted, the realities of the market are that the actual differences that would be exhibited between valuers using this technique may not be any greater than those produced using the traditional approach.

In the light of this, perhaps heretical, view, it will be useful to review the characteristics and possible sources of the three main variables in the model in slightly more detail.

The discount rate

The discount rate used in the valuation is clearly critical. Indeed, when used in a 'full' DCF it is far more critical than with the equated yield in the 'short-cut' approach. When the discount rate in the 'full' DCF is varied, then the NPV will change. With the 'short-cut' DCF, changing the discount rate will alter the implied annual growth rate in the calculation, either up or down depending on the direction of change. This will have a neutral effect on the actual valuation.

Traditionally, where the actual equated yield or discount rate or target rate (in this context, the three are essentially the same thing – discount rate will be used for simplicity when the 'full' DCF approach is referred to) are unknown, it is estimated relative to the rate of return on long-term government fixed-interest securities (gilts). Usually it is taken that property should attract a premium over long-term gilts due to increased risk, illiquidity and management cost. The most common premium added is 2% above the long-term gilt rate, so if gilts yield 8%, property should return at least 10% to persuade investors to buy it in preference. In fact, the risk premium will depend upon the type of property being considered – ie how risky is it? Is it a very illiquid investment or one that is easily sold? Is it costly or cheap to manage? This will vary from property to property and will vary over time according to market conditions. The premium on prime properties could vary between 1% and 5%; on secondary properties the premium could be higher.

The selection of discount rate can be made subjectively based upon the principles outlined above, or can be derived in other, more analytical ways. These include:

- use of the Capital Asset Pricing Model (see Brown, 1991, and below);
- analysis of the performance of similar investment properties;
- surveys of the requirements of investors in the market for the type of property being valued

All of these methods have their problems. The first two ideally require large data sets on the performance of properties of a similar type over a sufficient time period, something that, in the past at least, was often not available. Also, there is always the question whether patterns in the past are relevant to the future. The last approach requires investors to be honest, open and cooperative. Again, these characteristics are not always evident in the property investment markets.

In many cases, however, the problem of discount rate selection is not one that the valuer has to face. In many of the larger firms in the City of London discount rates are selected by specialists within the investment departments.

Rental growth rates

Where implied growth rates are not being used, rental growth rates must be estimated, assumed or forecast. As illustrated in the valuation examples above, either constant average rates or modelled rates of growth can be used. The key problems with rental growth in 'full' DCF valuations are firstly that the models are often extremely sensitive to changes in the rates assumed and also that rental growth in most rental markets is quite volatile. Much effort has been expended over the last decade to improve rental forecasting with some success, at least in terms of short-term forecasting, but the problem still remains. Nathakumaran and MacLeary (1988) suggested that it might be possible to use short-term explicit forecasts of growth and longer-term implied rates, and logically there is much merit in this suggestion.

Other dangers exist with the 'full' DCF approach. Unlike the 'short-cut' DCF and the arbitrage pricing model, they are sensitive to the time period of appraisal. At any point in time in the 'short-cut' DCF the valuation will be the same. This is not the case with the 'full' DCF because the relationship between the all-risks yield, the equated yield and the rental growth is broken.

The sale or terminal yield

Whereas in the 'short-cut' DCF model the yield used to terminate the explicit part of the cashflow is the current rack-rented ARY, with the 'full' DCF model the yield used at this point may be explicitly assumed. The most logical way of arriving at this yield is to analyse the performance of properties that are essentially similar to the subject property but which are older and have recently been sold. With a large enough sample, the depreciation characteristics of the investment concerned should be apparent and a reasonable estimate made.

There are two main potential problems with this. Firstly, this assumes that market conditions in the future will be similar to those that exist today. Yields on any given type of investment are children of their time. They depend on how investors view the relative merits of investments at the point in time when the investment decision is made. Over time yields vary, as investors' views of the merits of other investments and their views of the future also vary. Secondly, the yield profile assumes that depreciation characteristics in the future will be similar to those which have occurred in the past. This may not be the case: as we have seen, technology and fashions change very quickly. Depreciation may be faster or slower than the past, depending on changes in what occupiers demand.

This review of the key factors is, perhaps, depressing and also seems to contradict what has been said above about the merits of this approach. However, all we have done here is to point out the pitfalls and problems with the approach. If valuers are aware of the dangers and limitations of the 'full' DCF approach then they are likely to be more careful in its application. The question is surely not whether the assumptions made in a valuation are accurate but whether they were reasonable ones to make at the time the valuation was made. This philosophy is essentially the one already operated by the courts in assessing whether a valuer has displayed negligence in current valuation practice.

Contemporary methods: review and conclusions

Contemporary methods of valuation, as do traditional methods, have both strengths and weaknesses.

The 'short-cut' DCF method has many advantages. It is the most mathematically correct of the methods used to date and puts the 'correct' weight on the term and reversionary components of the valuation. A major advantage is that the method reveals far more about the assumptions made by the valuer, although many factors still remain implicit.

A further characteristic is both an advantage and a potential disadvantage. This is that the technique remains essentially market-based. It requires that there are market transactions to determine the yields used in the reversionary part of the investment. The advantage to this over the more explicit DCF techniques is that the resulting valuation will be supported by the transaction evidence. The disadvantage is that the technique may be as exposed as the other techniques to situations where there is a lack of market evidence or where the property is complex or different from the market as a whole. Whether this is a valid observation will be explored below.

Complexity is what the 'full' DCF method deals with best, where there is limited market evidence and unusual future income patterns. However, there are problems with the effect of moving away from market evidence. In particular the method's sensitivity to variations in the inputs used should be noted as should its relative instability in outcome relative to the time period of analysis chosen. These disadvantages are so marked that it is recommended that the 'full' DCF approach should not be used on its own in a valuation, and should only be used where it is difficult to use the other methods.

Chapter 5

Potential or Advanced Income Valuation Techniques

This chapter considers methods and techniques that have yet to make the transition to widespread use in practice. It covers an eclectic range of techniques, many of which have their roots in pricing techniques used in the mainstream equity and gilt markets. There is insufficient room to cover all the techniques in great detail in this short text so we will concentrate on those methods considered to have the greatest potential for use in property valuation practice. Where other methods are mentioned, the reader has the opportunity to follow up the reading list at the end of the book to gain information on these areas.

Risk-explicit valuations and simulation

Explicitly taking risk into account in a valuation seems on the surface to be quite a major step forward but is, in fact, only a minor extension to existing techniques. Many commercial valuers will be familiar with the basic approach. Any valuer who has carried out a sensitivity analysis on a development appraisal has explicitly addressed uncertainty.

Certain valuation techniques naturally lend themselves to risk-explicit approaches. This is particularly true of the 'full' DCF method where the valuer has to forecast future cash flows. One of the criticisms that can be (and is) levelled at this approach is that the forecast made by the valuer is bound to be uncertain, ie it may be found to be wrong. Just as an exploration of the sensitivity of the assumptions of a development appraisal is generally taken to be essential, addressing the uncertainty of the predictions made in a 'full' DCF valuation should also be viewed in the same way. There are various ways of addressing risk, some of which are relatively simple while others are more complex. An example of the latter is Monte Carlo simulation, an approach that is simple in concept but until recently was very time-consuming to carry out. This is one

Table 5.1 Probability table

Scenario	NPV £	Probability (p)	P × NPV £
Base	1,453,765	0.5	726,882.50
Low growth	1,182,396	0.25	295,599.00
High growth	1,612,145	0.25	403,036.25
Expected net present value			1,425,517.75

area where technical advances have brought this very powerful technique within easy reach of the majority of commercial valuers.

First, let us consider simple ways of taking risk into account and incorporating these assumptions into the final valuation. Taking the base property considered in Chapter 4 when discussing the 'full' DCF model, it should be obvious that the assumptions made in this valuation are only one possible outcome of the future. It is likely to be the valuer's 'best' or most reasonable assumption of what the future holds but if the economy is either better or worse than the valuer expects then either an over- or under-valuation may result. The most significant effects of these two alternative scenarios would be on the overall growth in rental values over the valuation period and on the terminal or sale yield on the property.

The original, most likely outcome is presented in Valuation 5.1, as are two alternative views of the future. One assumes that the economy will perform weakly over the valuation period (or alternatively that the property will perform more weakly that expected). The other assumes that the property will perform better than expected or that the economy will be stronger than anticipated. This is in itself a useful exercise but it can be taken a stage further by making an assessment of the likelihood of the alternative futures occurring. Let us assume that the valuer here decides that a reasonable assumption is that the original assumption has a 1 in 2 (ie a probability of 0.5) chance of occurring while the alternatives have a 1 in 4 likelihood (probabilities of 0.25). These assumptions can be incorporated into the valuation by constructing a probability table as shown in Table 5.1.

The valuation in Table 5.1 is slightly lower than that initially calculated suggesting that the downside risk had been under-estimated, at least according to the assumptions made.

Table 5.2 Possible alternative values for annual rental growth

Possible values for annual rental growth %	Probability of occurrence %	Ascribed probability numbers
0	2	1–2
1	5	3–7
2	7	8–15
3	10	16–25
4	15	26–40
5	21	41–61
6	15	62–76
7	10	77–86
8	7	87–93
9	5	94–98
10	2	99–100

Although this does represent an advance on the single point estimate, particularly for a DCF valuation, the scenario and probability approach is still fairly limited. Only three alternative views of the future are examined, each of which are discrete value assessments. Even taking the limited number of variables we chose there are an almost infinite number of alternative combinations of values, each of which can produce a distinctly different NPV calculation. To carry out individual calculations for each of the possible alternatives, having decided on its probability of occurrence, is possible in theory but impractical.

There is an approach that does move a long way towards examining a larger sample of the possible outcomes. This is Monte Carlo simulation. Simulation has been used since the 1960s to explore many areas of uncertainty, although until recently it has been a cumbersome method. It requires the range of probable values for a variable to be identified, for probability to be ascribed to each and then for random numbers to be generated in order for the value to be used in the calculation to be selected. Taking our example as an illustration, the possible alternative values for rental growth over the holding period and sale yields are as shown in Tables 5.2 and 5.3 respectively.

With our two-variable model two sets of random numbers, between 1 and 100, would be generated, one for each distribution. Let us say these are 22 and 67. For rental growth this would equate

Valuation 5.1 Probability weighted scenario approach using a 'full' DCF model

Scenario 1: base assumption/most likely forecast of cash flows

Rent	£100,000	Discount rate	10.00%
RR yield	7%	Forecast annual	
Reversion	3	growth rate	5.0000%
OMRV	£110,000	Terminal yield	9%

Year	OMRV £	Lease rent £	Sale value £	CF £	DF	DCF £
0	110,000					
1	115,500	100,000		100,000	0.9091	90,909
2	121,275	100,000		100,000	0.8264	82,645
3	127,339	100,000		100,000	0.7513	75,131
4	133,706	127,339		127,339	0.6830	86,974
5	140,391	127,339		127,339	0.6209	79,067
6	147,411	127,339		127,339	0.5645	71,879
7	154,781	127,339		127,339	0.5132	65,345
8	162,520	127,339	1,805,779	1,933,118	0.4665	901,814
					NPV	**£1,453,765**

Scenario 2: assumption of low growth/weak economy over analysis period

Rent	£100,000	Discount rate	10.00%
RR yield	7%	Forecast annual	
Reversion	3	growth rate	2.0000%
OMRV	£110,000	Terminal yield	10%

Year	OMRV £	Lease rent £	Sale value £	CF £	DF	DCF £
0	110,000					
1	112,200	100,000		100,000	0.9091	90,909
2	114,444	100,000		100,000	0.8264	82,645
3	116,733	100,000		100,000	0.7513	75,131
4	119,068	116,733		116,733	0.6830	79,730
5	121,449	116,733		116,733	0.6209	72,482
6	123,878	116,733		116,733	0.5645	65,893
7	126,355	116,733		116,733	0.5132	59,902
8	128,883	116,733	1,288,825	1,405,558	0.4665	655,703
					NPV	**£1,182,396**

Valuation 5.1 contd

Scenario 3: assumption of high growth/strong economy over analysis period

Rent	£100,000	Discount rate	10.00%
RR yield	7%	Forecast annual	
Reversion	3	growth rate	7.0000%
OMRV	£110,000	Terminal yield	9%

Year	OMRV £	Lease rent £	Sale value £	CF £	DF	DCF £
0	110,000					
1	117,700	100,000		100,000	0.9091	90,909
2	125,939	100,000		100,000	0.8264	82,645
3	134,755	100,000		100,000	0.7513	75,131
4	144,188	134,755		134,755	0.6830	92,039
5	154,281	134,755		134,755	0.6209	83,672
6	165,080	134,755		134,755	0.5645	76,066
7	176,636	134,755		134,755	0.5132	69,150
8	189,000	134,755	2,100,005	2,234,760	0.4665	1,042,532
					NPV	**£1,612,145**

Table 5.3 Possible alternative values for terminal (sale) yield

Possible values for terminal (sale) yield %	Probability of occurrence %	Ascribed probability numbers
7.75	1	1
8.00	4	2–5
8.25	7	6–12
8.50	10	13–22
8.75	15	23–37
9.00	21	38–58
9.25	15	59–73
9.50	10	74–83
9.75	7	84–90
10.00	5	91–95
10.25	2	96–97
10.50	1	98
10.75	1	99
11.00	1	100

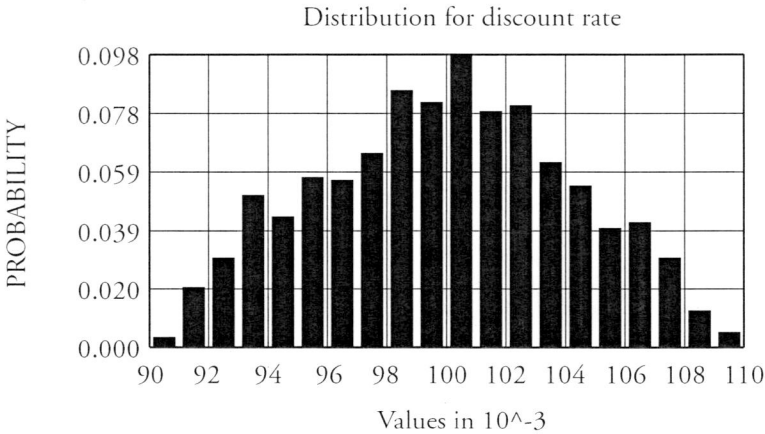

Figure 5.1 Discount rate probability distribution for simulation (figure of 100 is equivalent of 10%)

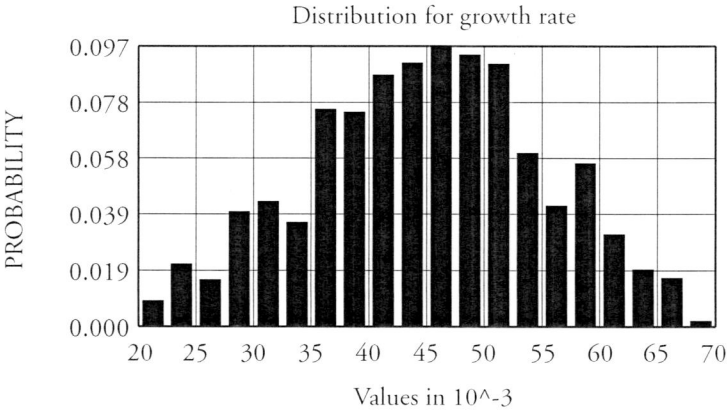

Figure 5.2 Rental growth rate probability distribution for simulation (figure of 50 is equivalent of 5%)

to 3% annual growth, while for sale yield the appropriate figure would be 9.25%. Combining these in the cash flow would produce an NPV of £1,293,784. To produce the final answer this process would have to be repeated at least 100 times which would give an array of NPVs which could then be analysed. In particular the mean value is important as this would be the figure used in the report, but the distribution of the NPVs would give a valuable insight as to the certainty or reliability of that figure.

The big problem with this approach (leaving to one side the problem of forecasting and probability assessment which will be addressed later) is clearly how time-consuming it is. It is possible to construct Excel spreadsheets incorporating macro functions that will take some of the drudgery out of the task but this does require a relatively high level of spreadsheet knowledge and skill. Fortunately there is software available as an add-in to spreadsheets like Excel at reasonable costs that allow simulations to be carried out very easily. One of the best examples is Pallisade Corporation's @Risk™ software. @Risk allows any input cell in a spreadsheet to be expressed as a probability distribution and will use these to quickly recalculate the spreadsheet and then analyse the results on any selected output cell. In this case we would clearly select the NPV calculation as the output cell.

An @Risk analysis was carried out on our example appraisal. The two variables selected as a probability distribution were rental growth and terminal yield but any number could have been chosen – the software allows you to choose from a wide array of possible distributions. For this example, a simple triangle distribution was chosen for each for which the minimum, maximum and most likely values are required, the software then calculating the probability distribution. These are displayed as Figures 5.1 and 5.2.

The simulation can then be run, selecting the NPV calculation cell as the @Risk output. In this case 1,000 iterations or individual calculations of NPV were carried out. On a relatively modestly powerful PC this process takes 5–10 seconds. The software reports the mean NPV plus a wide range of other statistics including the standard deviation. In this case the mean valuation was £1,443,551 with a standard deviation of £158,231. The distribution of the NPV calculations produced by @Risk are displayed in Figure 5.3.

This section of the text is not intended as a commercial for @Risk. There is no doubt, however, that this software greatly increases the ability of the valuer to carry out very sophisticated tasks. What does the use of simulation add here to our valuation? First, we have explicitly addressed the uncertainty contained within the valuation. Secondly, we have revealed more about the reliability of this valuation based upon our assessment of future uncertainty.

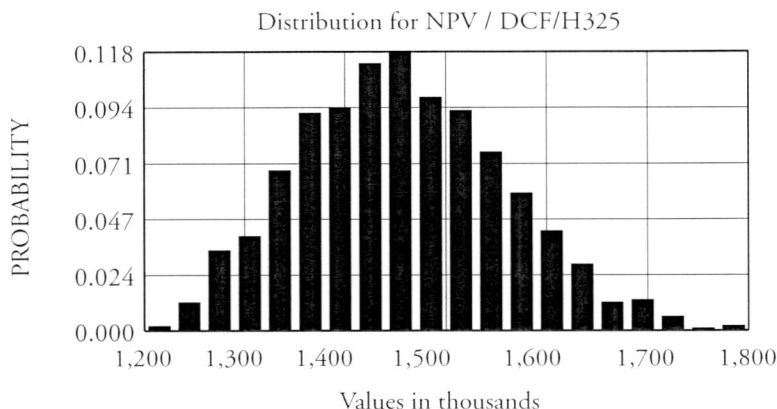

Figure 5.3 NPV output distribution representing 1,000 calculations during simulation

The valuation and standard deviation calculation enables us to calculate the coefficient of variation (the mean divided by the standard deviation expressed as a percentage). Here the coefficient of variance is 10.961%, meaning that just over 68% of the value estimates, based upon the probability as assessed, fall within about ±11% of the mean figure reported. This can be used as a benchmark to compare the reliability of the valuation in comparison with the valuations of other properties. For example, if most simulations were producing coefficients of variation below this level then the commissioner of this valuation would be able to gauge that the value of this property was subject to a relatively higher level of uncertainty. This type of information would assist lenders on property, in particular, in arriving at more rational decisions.

There are of course problems with this approach. Although this extension of the DCF technique counters some of the problems with the method that were identified above, namely the problems of accurate forecasting, other problems are peculiar to risk-explicit approaches. The principal problem is assessing probabilities. There is insufficient reliable data available to assess the likelihood of certain factors occurring in the future. Extrapolating from past performance and relationships may indeed not be a good way of assessing what will happen in the future. In essence, although the knowledge and experience of the valuer is helpful in assessing what the near future is going to be like, no one has perfect foresight. The assessment of probabilities is likely to be subjective.

What is the effect of changing, say, the range of estimates of the key variables? If the range of possible rental growth rates is extended to –2% and +12% with the same most likely figure of +5%, the mean valuation of the property after the simulation hardly changes (to £1,452,727) but the variability is higher (a standard deviation of £219,796). Here the coefficient of variance is 15.13%, ie suggesting that the valuation is less reliable. This does undermine some of the qualities identified above.

Despite these drawbacks, simulation is a very powerful and valuable technique. Just as DCF requires valuers to be explicit about what assumptions they have made about the future performance of the key variables, so simulation requires valuers to be explicit about the reliability of these forecasts. It is a logical extension of DCF that adds much to the single point estimate of NPV.

Valuation by arbitrage

The arbitrage approach to valuation is a relatively newly developed technique although the fundamentals that underlie it are long established from financial theory, from where the method is derived. It was formulated by French and Ward (1995) and subsequently developed further by French and Ward (1996) and French and Ward (1997). Although the basic concept may be considered to be complex, in practice an arbitrage valuation is relatively easy to carry out.

Arbitrage is based in the premise that any asset, including cash flows, should be valued by comparison with other assets with similar risk characteristics. In financial markets, dealers practise arbitrage regularly, investing in one asset but laying off or covering the risk by taking a contrary view elsewhere. The futures and options markets are classic examples of such use.

It is perhaps easiest to illustrate the principles by looking at our now familiar base property. In risk terms it consists of two parts: the term, which is low risk as the income is secure and known until the next rent review, and the reversion, where the rent is unknown at the time of the valuation. The risk related to the reversion is, thus, higher. It seems logical that this higher risk be reflected in the yields used to value the two types of income stream. This is the principle behind the manipulation of yields in the split yield traditional models. The difference with the arbitrage approach is that the approach is mathematically and logically defensible.

**Valuation 5.2 Implicit arbitrage valuation of the subject property –
analysis of comparable**

Assumptions/facts

Rent	£100,000	Low risk yield	8.50%
RR yield	7%	DCY	?
Reversion	5		
OMRV	£100,000		

Calculation

Term		£100,000	
YP 5 yrs @ 8.50%		3.9406	£394,064
Reversion		£100,000	
YP perp. @ 7.00%	14.286		
Def. 5 yrs @ ?	?	?	£1,034,507
			£1,428,571

Using the goal seek function on an Excel spreadsheet it is relatively easy
to calculate this discount rate. The solution is illustrated below.

Assumptions/facts

Rent	£100,000	Low risk yield	8.50%
RR yield	7%	DCY	6.67%
Reversion	5		
OMRV	£100,000		

Calculation

Term		£100,000	
YP 5 yrs @ 8.50%		3.9406	£394,064
Reversion		£100,000	
YP perp. @ 7.00%	14.286		
Def. 3 yrs @ 6.67%	0.7242	10.345	£1,034,507
			£1,428,571

What yields should be used for each part? French and Ward (1995)
look at the price that should be paid from the viewpoint of the
tenant. The rent on the term is a known outgoing, fixed until the
next rent review. It is not dissimilar to a fixed interest loan,
therefore the yield used should be equivalent to the firm's cost of
borrowing over the same period. We will assume that the cost of
borrowing to the firm is 8.5% in this case.

The yield for the period after the rent review is slightly more difficult. Arbitrage principles suggest that the tenant should act to cover the liability by investing in an asset with the same risk characteristics that should grow to cover the future liability. As we are considering the rent of an office property, it is logical then that, notionally at least, the tenant should invest in the freehold interest of an exactly similar property which should grow at the same rate as the subject and thus cover the liability in the future. This sounds complex, but in reality it is not. It merely means that the value of the reversion is derived from the analysis of a rack-rented freehold of the same type.

The process is illustrated in Valuation 5.2. A rack-rented property, similar to the subject in all other ways, has been sold at a price reflecting an initial yield of 7%. Using this transaction but following the arbitrage principles, the transaction can be analysed in a different way. Valuing the term using the finance rate of 8.5% gives the 'correct' valuation of the reversion. Knowing what the sale price is and also that the reversion will be worth the 7% initial yield to investors at this point in the market, it is possible to calculate the appropriate discount rate for this, risky, part of the income flow. This can either be done by formula or else by trial and error using a term and reversion type approach. This is illustrated in Valuation 5.2. This discount rate, called the discounted capital yield by French and Ward (1995), seems rather low to the casual observer. In fact it contains, implicitly, allowances for growth and for the uncertainty concerned. This yield can then be applied in the valuation of our subject property (Valuation 5.3).

It is possible to produce a growth-explicit version of the same type of approach. The discounted capital yield can be converted to the capital yield using the following formula:

$$\text{CY} = [(1 + \text{DCY}) \times (1 + g)] - 1$$
$$= [(1.0667) \times (1.034209)] - 1$$
$$= 10.32\%$$

This yield is then used in the valuation as illustrated in Valuation 5.4. As can be observed, this produces a valuation that is identical to the non-growth explicit approach.

The advantages and disadvantages of the arbitrage approach are similar to those for the 'short-cut' DCF approach, with the added advantage that a mathematically defensible valuation can be produced without explicitly calculating the growth rate. The technique

---

Valuation 5.3 Implicit arbitrage valuation of subject

Assumptions/facts

Rent	£100,000	Low risk yield	8.50%
RR yield	7%	DCY	6.67%
Reversion	3		
OMRV	£110,000		

Calculation

Term		£100,000	
YP 3 yrs @ 8.50%		2.554	£255,402
Reversion		£110,000	
YP perp. @ 7.00%	14.286		
Def. 3 yrs @ 6.67%	0.8239	11.771	£1,294,773
			£1,550,175

Valuation 5.4 Growth-explicit arbitrage valuation

Assumptions/facts

Rent	£100,000	Capital yield	10.32%
RR yield	7%	Implied annual	
Reversion	3	growth rate	3.4209%
OMRV	£110,000	Low risk yield	8.50%

Calculation

Term		£100,000	
YP 3 yrs @ 8.50%		2.554022371	£255,402
Reversion		£121,679	
YP perp. @ 7.00%	14.28571429		
Def. 3 yrs @ 10.32%	0.744859557	10.64085082	£1,294,773
			£1,550,175

---

still requires, ideally, the presence of good comparable evidence, though this is not as marked where there is a substantial period to reversion as the adoption of the low risk discount rate does not require direct market evidence and represents a substantial proportion of the value.

Other possible techniques

This section covers techniques that are usually considerably outside the experience of practising investment valuers. A number of methods, drawn mainly from the financial and capital markets, are briefly examined. All of them have been proposed as valuation methods and many of them appear to be very promising in their ability to produce rational prices for property. As yet, however, none have been widely adopted by the industry.

Asset pricing models

There are a number of models that arise from the world of capital pricing in the financial markets. One of the oldest, and the one which has generated the most debate about its validity but which is still very influential, is the Capital Asset Pricing Model (CAPM). Another, arbitrage price theory, forms the basis of the arbitrage approach, but is a complex set of theories in its own right. Because of its basic simplicity and potential value, only CAPM will be reviewed in detail here.

CAPM is not, in itself, a valuation technique in the mode, say, of the 'short-cut' DCF method. Its main purpose is to determine an appropriate rate of return, or equated yield, for any particular investment. This area is a particular weakness of the property market. While it is generally assumed that property should illustrate a 2% premium above that received on long-term risk-free investments, the equated yield for different classes of property and individual property investments may be more or less than this margin. The differences will largely arise from the different risk characteristics of each. It has been difficult for the market to rationally assess what the difference should be. CAPM is potentially of great assistance in this area in that it allows the calculation of the risk in an investment that cannot be diversified away by combining it in a portfolio. This risk – known as systematic risk – is viewed under the theory as the most important factor in determining the risk premium on any individual investment.

The calculation of the appropriate discount rate using CAPM requires three components:

- the rate of return on risk-free investments;
- the expected return on the market portfolio (or of an index);
- an assessment of the systematic risk of the investment.

This last element is expressed as the β (beta). The β is a measure of the relative variability of an asset compared with the market as a whole. It is usually calculated by regressing the returns from one asset against the returns from the market as a whole. The actual mechanics are relatively unimportant in this forum; what is important is the principle. The β is expressed as a numeric factor in comparison with the variability of the market which we can take to be 1.0. If the asset class has a β of, say, 0.8, it means that it is a little less variable in its return than the market as a whole. If the β is, say, 1.20, the asset class is slightly more variable in its returns than the market.

Using the CAPM formula the β can be used to calculate a rational rate of return for an investment class. The formula is:

$$E(r_i) = r_f + \beta_i [E(r_m) - r_f]$$

where

$E(r_i)$ =   expected return on investment
β_i =   beta of investment asset class
r_f =   rate of return on risk-free investments
$E(r_m)$ =   expected return on the market as a whole.

The best way to illustrate the principles of this is to use an example. Let us assume we are trying to value an office in Edinburgh. We have collected the following data:

• the rate of return on 'risk-free' investments is 8%;
• the expected return on the market (eg the IPD index) is 10%;
• the β on Scottish offices is 1.5.

It should be noted that these figures are, of course, hypothetical.
Inserting these figures into the formula gives us:

$E(r_i)$ = 8% + 1.5 [10% – 8%]
       = 8% + 3%
       = 11%

This is a very powerful and persuasive technique, although it has been criticised for its over-simplification of the relationship. There are other problems in applying this technique – which, remember, comes from the equity and gilts markets – to property. There are real problems in determining the β. First, although there are now indices of investment property returns to use in the analysis, largely thanks to the Investment Property Databank (IPD), these indices have been criticised by some academics for being determined by valuation.

Valuations of properties in portfolios suffer from 'smoothing', ie the previous valuation influences the next, which means that the actual degree of variance may be understated in the index. Other problems include the reliance on past relationships and returns to determine the future. In fact these relationships may not be valid in the future. Another problem is that the technique requires the assumption of the basic risk premium that property should show over the risk-free rate. Although it is assumed to be 2% it is not known securely.

Despite this, CAPM offers one of the few rational ways of determining an appropriate discount rate and has much potential for use in valuation.

Contingent price models

The origins of contingent price models are rooted firmly in the financial markets. Strictly speaking we have already looked at an example of this approach in the arbitrage method of valuation. We have returned to the principles of the approach again because, in addition to arbitrage, there is a range of other contingent price models that many feel have applications in property valuation, particularly in complex situations. In particular there are models used to value options in the financial markets that may be used to value things such as upward and downward rent review clauses or break options in leases.

Contingent price models are so named because they follow the principle that the value of any particular asset is contingent on the price of other, different, assets of similar characteristics. The components of the investment returns are divided up into parts with differing risk characteristics and are valued separately. This principle is used to value things such as insurance contracts, spread bets and financial derivatives like futures. The normal application of contingent valuation models is where there are options involved; for example, in a development project there may be options at various stages to proceed as planned, to postpone the project, to change the project or even to abandon it. These various options should be reflected in the price paid in the market and can be used to value other similar projects.

As the roots of these projects are from the financial markets it is often easiest to explain the principles using a financial market example and then explore how they can be applied to property. This is the approach that will be adopted here. We will look at two models in particular, arbitrage and risk-neutral approaches to valuation. The reader should be aware that there are other models and also that the simulation or Monte Carlo modelling technique we have already examined is a flexible alternative.

Arbitrage

The way arbitrage is operated in the financial markets is to find two forms of what are essentially the same asset, trade them and profit from any differences in pricing. Nick Leeson was practising arbitrage in Singapore, trading in the same assets, futures contracts, in Japan and Singapore. The Japanese market was much larger and more sophisticated than the Singapore futures market which still set prices by 'open outcry', ie traders meeting face to face on the trading floor. As the Japanese market traded small price differences could open up between the two markets, and buying in one with a balancing trade in the other allowed profits to be made. It should have worked – and indeed it did. Unfortunately, Mr Leeson's behaviour and the structure of Barings allowed things to go rather wrong!

These basic principles can be used in another way. Investors construct and trade portfolios that behave like other assets. An example of this is the relationship between an option price on an index of shares and the price of the component shares in that index. This is also referred to as arbitrage. This can in turn be used in valuation, using the value of known components of a portfolio to value other components whose cash flows are equivalent but whose price is unknown. This uses a principle called the Law of One Price, which states that assets that have the same value should have the same price.

An example of how this technique works can be seen from the following example (derived from RICS, 1998, Paper 3). A share has a price of 100p. Tomorrow it may rise to 200p or fall to 50p. In the portfolio is a bond also with a price of 100p, a price that will not change between now and tomorrow. There is also an option which pays 300p if the share price rises but nothing if it falls. The value of this option is unknown. In order to value the option a portfolio is created consisting of 2 shares bought and 1 bond borrowed. The return from the bond is assumed to be zero in real terms. The cost of the portfolio is 200p – 100p = 100p. If shares rise to 200p it will be worth 400p – 100p = 300p. If it falls to 50p it will be worth 100p – 100p = 0p. This can be seen to be exactly the same cash flow potential as the option. The option must be worth 100p also, ie the same price as the notional portfolio.

This basic principle is what underlies the arbitrage approach to valuation considered earlier. French and Ward (1996) have applied the method to valuing upwards-only rent review clauses. It clearly has potential for property valuation though it may require the market to start making the same assumptions about valuing different options in leases and investments before it becomes a good predictor of prices in the marketplace.

Risk-neutral approaches to valuation

Risk-neutral models are an alternative approach to arbitrage. The approach assumes an environment in which all investors are risk-neutral and therefore all assets would be expected to give the same return. Once an asset has been valued in this way the cash flows can be divided up into risk-neutral components and each valued separately. This is rather hard to follow, so to aid understanding we will consider an example to explain how it actually works. The following example is derived from Paper 3 of the RICS's report *Right Space: Right Price?*, published in March 1998, which discusses many of these financial models in the context of property valuation.

Imagine that we are considering a share that is currently priced at 100p. We feel that in one year's time it might rise to 200p or fall to 50p. There is an option available that will pay 300p if the share rises but nothing if it falls. We will also assume that the risk-free rate is 0% and that the expected return on the shares is 20%.

We can calculate that the probability of the shares rising as follows:

Current market price = (Value if shares rise × Probability of rise + Value if shares fall × Probability of fall) /1 + Expected return

In this case this is:

100 = (200 × Probability of rise + 50 × Probability of fall) / 1.20

This equation is relatively easy to solve, particularly using the goal seek function of Excel.

$$100 = (200 \times 0.467 + 50 \times 0.533) / 1.20$$

What we need to do now is price the option. The option is riskier than the share but we don't actually know how much riskier. What we need to do is follow the principle of pricing in a risk-free world. The current value of the share reflects the discounted value of the future cash flows, which are the share prices in this example. If the risk-free rate is zero, that forecast would be the same as today's

Figure 5.4 Risk-neutral approach to valuing an upwards-only rent review premium

		Values	Years 1–5	Years 5–10 (up)	Years 5–10 down
Upwards-only lease	Market rents		100,000	120,000	95,000
	Formula		100,000 × YP 5 years at 7.5%	120,000 × YP 5 years at 7.5% × PV 5 years at 7.5% × Probup	100,000 × YP 5 years at 7.5% × PV 5 years at 7.5% × Probdown
		700,000	404,588	81,556	213,857
	Yields	7.50%			
	Rent review period	5			
	Risk-neutral probabilities			0.241157651	0.75884247474
Up or down lease	Market rents		100,000	120,000	95,000
	Formula		100,000 × YP 5 years at 7.5%	120,000 × YP 5 years at 7.5% × PV 5 years at 7.5% × Probup	95,000 × YP 5 years at 7.5% × PV 5 years at 7.5% × Probdown
		689,308	404,588	81,556	203,164
	Yields	7.50%			
	Rent review period	5			

price, ie 100p. In fact we know them to be 200p or 50p. Following the same approach as above, but assuming a risk-free return, the formula below is used to determine the probabilities of a rise or fall in this risk-free world:

$$100 = (200 \times \text{Probability of rise} + 50 \times (1 - \text{Probability of rise})) \, / \, 1.00$$
$$= (200 \times 0.333 + 50 \times 0.667) \, / \, 1.00$$

The probability of the shares rising is therefore 33.33%. We can now use this to value the option as we can discount the expected value at the same risk-free rate:

$$\text{Value} = \quad 0.3333 \times 300 \, / \, 1.00$$
$$= 100p$$

Note that the value of the risk-free rate need not be reduced to zero. This has been done for simplicity. If the risk free-rate had been 10%, for example, the probability of rise would have been 40%. The value would then have been 109.09p.

This method is a very powerful way of analysing the pricing of assets and it can be applied to property valuation. The RICS report gives an example of the valuation of an upwards-only rent review premium, something the risk-neutral approach is well suited to tackling.

This is illustrated in Figure 5.4. The example used is a ten-year lease that has been valued at £700,000. This is a conventional UK lease in that it contains an upwards-only rent review clause. The lease is used as a base for the calculation of the cash sum that should be paid to a landlord as compensation for giving up the upwards-only clause. It is estimated that market rents may rise to £120,000 pa or fall to £95,000. The risk-free rate is 7.5%. The procedure involves calculating the risk-neutral probabilities of the scenarios occurring and then applying these to the cash flows of the upward and downward lease. In the latter of course the rent will fall to the market rent at the time of the rent review, thus realising a fall in values. As can be observed this produces a difference of £10,693.

Although this is quite a convincing example there are some questions raised by the approach. First, the example is rather simplistic in that only the two different rent levels on review are considered. The authors of the RICS report recognise this, noting that other rental levels could be incorporated but this would greatly increase the complexity of the calculation. Another criticism is that the approach may have more difficulty with other types of property investments. With the example chosen for evaluation, the

upwards-only clause, it is relatively easy to conceptualise the
difference between the two forms. With something like a lease with
a break clause and a lease without a break clause, this comparative
conceptualisation becomes much more difficult. Even in the two
rental scenarios used in the above example a break may or may not
be operated. Tenants make decisions due to factors other than
purely the level of rent. In practice it would be much harder to
decide whether the risk-free probabilities calculated in one situation
would be valid in another.

It is accepted that these problems are not technically insurmount-
able but they add to the complexity of using the technique to such
an extent that it must be questionable whether this method is
superior to, say, a Monte Carlo simulation approach.

Review and conclusions

There can be little doubt that these asset pricing models do have the
potential for providing assistance to the valuer. This potential
seems to be greatest where the valuer is facing situations that the
conventional models have difficulty in tackling. The techniques
have mathematical and logical validity.

However, there can also be little doubt that the methods have
problems in their application to property valuation, some of which
have been mentioned above. Fundamentally, it should be
recognised that the methods were derived from fast-moving, data-
rich environments where the players in the market tend to behave
rationally. These are not the characteristics that typify the property
investment market. For all their undoubted qualities this is a factor
which may well restrict the adoption of these techniques.

The RICS (1998, Paper 3), as referred to above, recognises that
research in this area is at an early stage for property applications. It
also notes that valuers and property investors will need to develop
knowledge and skills in these areas at some point in the future in
order to compete ahead of the competition. The big question is
when this should be. Are we at that stage now? This is a very
difficult question to answer. If the question is rephrased as 'can the
majority of valuers of investment property carry out their tasks in
the short to medium term without applying these methods?' then
the answer to this is, at this stage, probably yes. As we have seen,
however, changes in the property investment market seem to
accelerate over time. The valuation profession and property
investors need to constantly monitor and review the situation.

Conclusions to Part 1

In this part we have explored some of the traditional, some of the contemporary, some of the advanced or potential and some of the possible methods of valuation that can be used to value property investments. However, it should be noted that there are yet further methods, for example the 'real value' approach (Wood, 1972), that have not even been mentioned, while there are also numerous variants on the approaches we have looked at. All this is indicative of the fact that no one method is the ideal; each of the approaches we have looked at has strengths and each has weaknesses. These strengths and weaknesses depend a lot on the circumstances in which the valuation is being carried out.

In a simple market (if there is such a thing) with plenty of evidence, the traditional approaches may well be best at producing a prediction of a transaction price, which is what most valuation is aimed at doing. It will do so largely because the people agreeing the sale and purchase will be using those self same methods. It is often forgotten that in the commercial investment market, unlike the residential market, we are not observing the behaviour of independent buyers and sellers but instead the results of pairs of valuers/surveyors carrying out appraisals and valuations. If the market sets its transaction price using traditional, non-rational methods then there is little point using rational contemporary models to produce a valuation, for the answer produced may well be different.

In more sophisticated or complex circumstances things are likely to be different. As such conditions become more common, so the methods used in the market will change. This will be something that is learned by the market as it is realised that the traditional or familiar methods cannot cope and may produce wildly divergent figures from others in the market. It seems likely that the methods used in valuation will not change completely but that coexistence of advanced and traditional approaches will be observed.

Given this view of valuation, it seems logical to explore how the various methods perform in some of the more 'stressful' situations for the traditional methods. By doing this the practitioner will become more aware of where there is a greater risk that their preferred model will fail. All the approaches, except those covered in the section on 'possible' valuation methods in Chapter 5,[1] will be tried under various circumstances to see how robust and reliable they are. The main approach to this will be to assess how sensitive the methods are to differences in assumption, ie how easy it is to produce one valuation that differs markedly from another where the evidence is unclear. After this has been done, the performance of the various methods will be evaluated.

Note

1   The reason for leaving out the 'possible' valuation methods is that the academic world has not fully evaluated these approaches. Their use seems to be too remote for the vast majority of practitioners to include them within Part 2. If the pace of market changes continues to increase, however, this view may well have to be revised in the not too distant future!

PART 2

Applying the Toolkit to Real-World Situations

This book is intended to be largely practical, rather than theoretical, and will now explore how the various methods deal with 'real-world' problems facing the valuer. How good is each method? How well do they deal with each circumstance? Which is best in what situation?

This part looks at three problems in capital valuations:

- short leases;
- break clauses in leases;
- over-rented property investments.

In turn the problems that each pose for valuers will be examined, then each approach will be applied followed by a critical appraisal of the result.

Chapter 6

Short Leases

Problems posed for valuers

As noted above, the UK market has, in the past, provided investors with long leases giving them substantial security of income. The commercial property crash of the early 1990s allowed tenants to negotiate leases that offered them greater flexibility, particularly in the office market. This was exhibited in two main ways: shorter leases and the inclusion of break clauses in leases. Although both have the effect of lowering the length of occupation and increasing the uncertainty for investors, the problems for valuers are slightly different and will be considered separately here.

With short leases, the investor loses the security that potential breaks in the income flow will occur so far in the future that they will have relatively little impact on value. The effect, of course, is due to the time value of money and can be illustrated by a simple example. Both of the investments in Figure 6.1 have the same nominal value, 18 individual receipts of £1,000. Investment B has an interruption in the income flow in years 6 and 7, whereas for investment A this does not occur until years 19 and 20. The net effect on the value is, however, a difference of £765.45.

Shorter leases, with the possibility of frequent interruptions in income flow that are earlier than expected, have a disproportionate effect on value. A further complication is that the interruption is not certain – a tenant may renew the lease. Even in this case there is more uncertainty – the end of the original lease may give the tenant an opportunity to negotiate new lease terms. The tenant in this case may well be in a stronger situation than where there is a rent review. If the tenant does not renew, there are questions of how long the void period will be. What quality of tenant will take the new lease and at what price?

INVESTMENT A				INVESTMENT B			
Discount rate: 10%				Discount rate: 10%			
Year	*C/F* £	*D/F*	*DCF* £	*Year*	*C/F* £	*D/F*	*DCF* £
1	1,000.00	0.9090909	909.09	1	1,000.00	0.9090909	909.09
2	1,000.00	0.8264463	826.45	2	1,000.00	0.8264463	826.45
3	1,000.00	0.7513148	751.31	3	1,000.00	0.7513148	751.31
4	1,000.00	0.6830135	683.01	4	1,000.00	0.6830135	683.01
5	1,000.00	0.6209213	620.92	5	1,000.00	0.6209213	620.92
6	1,000.00	0.5644739	564.47	6	–	0.5644739	–
7	1,000.00	0.5131581	513.16	7	–	0.5131581	–
8	1,000.00	0.4665074	466.51	8	1,000.00	0.4665074	466.51
9	1,000.00	0.4240976	424.10	9	1,000.00	0.4240976	424.10
10	1,000.00	0.3855433	385.54	10	1,000.00	0.3855433	385.54
11	1,000.00	0.3504939	350.49	11	1,000.00	0.3504939	350.49
12	1,000.00	0.3186308	318.63	12	1,000.00	0.3186308	318.63
13	1,000.00	0.2896644	289.66	13	1,000.00	0.2896644	289.66
14	1,000.00	0.2633313	263.33	14	1,000.00	0.2633313	263.33
15	1,000.00	0.239392	239.39	15	1,000.00	0.239392	239.39
16	1,000.00	0.2176291	217.63	16	1,000.00	0.2176291	217.63
17	1,000.00	0.1978447	197.84	17	1,000.00	0.1978447	197.84
18	1,000.00	0.1798588	179.86	18	1,000.00	0.1798588	179.86
19	–	0.163508	–	19	1,000.00	0.163508	163.51
20	–	0.1486436	–	20	1,000.00	0.1486436	148.64
NPV			**£8,201.41**	NPV			**£7,435.93**

Figure 6.1 Impact of the time value of money

To consider how the various approaches to valuation tackle this problem we will consider the following example:

A modern office property let two years ago on a five-year lease at a rent of £50,000 pa. The lease is on FRI terms, with no rent reviews, but is otherwise standard. The property's current OMRV is still £50,000.

Market evidence:
Rack-rented office investments let on 15-year FRI leases with five-yearly rent reviews to the then OMRV sell at prices equivalent to 7%.

Other information:
• Long-term government securities yield 8%.
• The normal risk premium for this type of investment is 2% above the long-term gilt rate.

In general, each of the valuations will be constructed based solely on this information, thus simulating the construction of a valuation in a 'thin' market. The strengths and weaknesses of each approach will be discussed, and limited sensitivity analysis of the majority of valuations produced will be conducted to examine the stability of the valuations produced by each approach. Essentially this will reproduce the effect of valuers making different assumptions about the key variables to simulate what may happen when different valuers are working with inadequate data, something that is relatively common. If an approach is sensitive to the constructional assumptions made then there is a greater risk of producing a valuation out of step with that which a 'reasonable' valuer might produce, ie producing a valuation that might be held to be negligent. The sensitivity analysis will give an indication of this. The analyses are not complete in that they do not examine the full range of possible variable assumptions. Instead individual variables will be adjusted in isolation, usually by altering the value by one whole unit of worth. Some of the key variables are also examined together in order to examine the effect on the valuation produced of different assumptions at the high or low end of the range.

Traditional approaches

The problem that traditional approaches suffer from in these circumstances relates to the availability of evidence. If there are enough transactions in the marketplace on shorter-term investments then it will be possible to make a reasoned judgement about how the investment market is appraising the risk profile of this type of investment. While this may occur, it is more likely that there will still be considerable doubt whether the subject and the transaction evidence are truly comparable. The real problem with these situations is that there are so many variables to consider that make each case individual. One five-year investment may have entirely different risk characteristics from another. If the market is diverse it is more difficult to make yield adjustments securely. There is still the benchmark, the ARY on the longer lease terms which provides a lower yield level, but adjustment above this level may be difficult.

There are a number of different ways of valuing these types of investment using the traditional approaches to income valuation. Three alternatives are:

• an initial yield approach;

Valuation 6.1 Initial yield approach

Assumptions/facts
Rent	£50,000
RR yield	7%
Reversion	3
OMRV	£50,000
Initial yield	10.00%

Calculation
Lease rent		£50,000	
YP perp. @ 10.00%		10	
			£500,000.00
Less costs		5%	£23,809.52
			£476,190.48

Valuation 6.2 Initial yield approach with an explicit void allowance

Assumptions/facts
Rent	£50,000
RR yield	7%
Reversion	3
OMRV	£50,000
Initial yield	9.00%
Void	2

Calculation
Lease rent		£50,000	
YP perp. @ 9.00%		11.111	£555,555.56
Less:			
Void allowance (2 years' rent)		−£50,000	
YP 2 yrs @ 9.00%	1.759		
Def. 3 yrs @ 9.00%	0.772	1.358	−£67,917.83
			£487,637.73
Less costs		5%	£23,220.84
			£464,416.88

- an initial yield approach with an explicit void allowance;
- a term and reversion approach with the reversion deferred to allow for a possible void period.

In all of these methods the yield is manipulated to allow for the increased risk involved with the investment. In the first approach, all of the allowance for future interruptions in income flow are allowed for in the yield. In the latter two, an explicit allowance is made for a possible void period.

Initial yield approach

The attraction of this approach is its simplicity and its direct relationship with market evidence. If there is sufficient market evidence to securely judge the correct market yield for the investment then it is the preferred approach. A problem with this method is that it does not reveal the assumptions that are contained within it about the void period etc, but this applies, to an extent, to all of the traditional methods. If the evidence is available then this method is the most reliable. If the evidence is not available, then the result can be sensitive to changes in the yield chosen, though as the yield gets higher, the effect of change (or error) becomes less. In the example shown in Valuation 6.1, it was assumed that the market evidence suggested a yield of 10%. Choosing a yield of 9% increases the valuation by 11.1%; choosing 11% lowers it by 9.1%. Despite this relative lack of sensitivity, this may mean that one valuer would value the investment at around £430,000 while another might value it at £530,000. This range might be difficult to defend in a negligence action.

An initial yield approach with an explicit void allowance

This is an alternative approach that some valuers might adopt. It actually produces identical results to an equivalent yield version of the third option. It offers the apparent advantage to valuers that an explicit assumption or allowance can be made for the potential void period. This is also an advantage to the procurer of the valuation who can more easily observe some of the key assumptions made. The valuation then resembles an equivalent yield valuation. The yield used would be less high than with the straight initial yield valuation. The approach is illustrated in Valuation 6.2.

Table 6.1 Sensitivity summary

	Current values	s1	s2	High comb.	Low comb.
Variables:					
Initial yield	9%	10%	9%	8%	10%
Void period	2	2	3	1	3
Result:					
Valuation	£464,416.88	£414,098.34	£436,023.20	£560,236.67	£387,218.63
Difference		−10.83%	−6.11%	20.63%	−16.62%

```
┌─────────────────────────────────────────────────────────────────────────┐
│ Valuation 6.3  Term and reversion equivalent yield                        │
│                                                                           │
│ Assumptions/facts                                                         │
│ Rent                      £50,000                                         │
│ RR yield                       7%                                         │
│ Reversion                       3                                         │
│ OMRV                      £50,000                                         │
│ Equivalent yield           9.00%                                          │
│ Void period (years)             2                                         │
│                                                                           │
│ Calculation                                                               │
│ Term                                          £50,000                     │
│ YP 3 yrs @ 9.00%                                2.531        £126,565     │
│ Reversion                                     £50,000                     │
│ YP perp. @ 9.00%          11.111                                          │
│ Def. 5 yrs @ 9.00%         0.6499               7.221        £361,073     │
│                                                                           │
│                                                             £487,638      │
│ Less costs                                         5%     £23,220.84      │
│                                                           £464,416.88     │
│                                                                           │
└─────────────────────────────────────────────────────────────────────────┘
```

The problem with this approach is that it is rather sensitive to the yield choice and the choice of void allowance. This can be observed from the sensitivity table provided in Table 6.1. The method is, ironically, much more reliant on good market evidence than the initial yield method, simply because the yield is lower. In addition, the extra variable of the void period assumption adds another area of potential variance. It is ironic because the reason this method might be used is where the market is diverse and there is not sufficient direct evidence to use an initial yield approach.

In summary the method can produce an acceptable result (taken to be one that the majority of valuers would produce given the evidence) but there are greater risks of variance.

A term and reversion approach with the reversion deferred to allow for a possible void period

Equivalent yield approach

The valuation can be set out in a term and reversion format. Where an equivalent yield approach is used then the valuation outcome

---

Valuation 6.4 Term and reversion split yield

Assumptions/facts

Rent	£50,000		
RR yield	7%		
Reversion	3		
OMRV	£50,000		
Reversion yield	10.00%		
Void period (years)	2		

Calculation

Term		£50,000	
YP 3 yrs @ 7.00%		2.624	£131,216
Reversion		£50,000	
YP perp. @ 10.00%	10.000		
Def. 5 yrs @ 7.00%	0.713	7.1298	£356,493
			£487,709
Less costs		5%	£23,224.23
			£464,484.66

will be identical to Valuation 6.2 (see Valuation 6.3). Essentially the calculation here is simply laid out in a different format but the basic calculation is the same. Given that the method also has the same strengths and weaknesses, it will display the same sensitivity.

Split yield term and reversion approach

It is also possible for a valuer to take a split yield approach to the problem. This is illustrated in Valuation 6.4. The philosophy behind this approach is that the valuer is trying to allow for the fact that the lease income is secure but that the future income is uncertain. In Part 1 of the book this method was criticised on a number of grounds but of the traditional methods it is actually the least sensitive to the individual inputs made (see Table 6.2). This is relatively surprising, though it should be noted that the combination of the three variables (term yield, reversionary yield and void allowance) do lead to high levels of variance in valuation when the extreme values are combined.

Table 6.2 Sensitivity summary

	Base values	s1	s2	s3	High comb.	Low comb.
Variables:						
Term yield	7%	6%	7%	7%	6%	8%
Reversionary yield	10.00%	10.00%	9.00%	10.00%	9.00%	11.00%
Void period	2	2	2	1	1	3
Result:						
Valuations	£464,484.66	£483,123.51	£502,208.80	£488,250.86	£546,383.46	£378,662.23
Variation from base valuation		4.01%	8.12%	5.12%	17.63%	−18.48%

Summary of the traditional approaches to the valuation of short leases

All of the traditional approaches share similar characteristics when asked to adapt to deal with these types of investments. The plus points are simplicity and familiarity. Valuers are used to using these methods and are likely to have a better feel for the manipulation of the yields and the assumptions required than with the contemporary approaches. This may seem to be a weak argument in defence of techniques that do have profound weaknesses and internal inconsistencies but there is little doubt that being familiar and comfortable with the method used is very important.

On the negative side, it is clear that when the valuer goes beyond the basic initial yield approach, the techniques are required to deal with situations that they were never designed for. It should be recalled that the approach developed at a time when income flows were long, secure and inflation free. They then adapted to deal with long income flows with inflation, though weaknesses were apparent. Now that they are being asked to deal with much shorter income flows and uncertainty about the future the adequacy of the methods really must be questioned. The traditional approaches rely on the implicit assumptions contained within the ARY. These situations really require the valuer to be explicit and address the uncertainty about the future.

Does this imply that the methods should not be used in practice to deal with the valuation of short leases? This is a difficult point to address. If there is sufficient evidence and the market is uniform then it may well be possible to derive sufficient evidence to determine the appropriate traditional yields. In addition, if the market is using these methods to determine the exchange price of this class of investment it is likely that using the same methods to value the property will be the best predictor of price in the market, which is, after all, the main function of valuation. To counter this, it is clear that markets are becoming much less uniform and investors are using alternative methods to appraise the worth of their holdings. In the light of this, and given the fundamental weaknesses of the methods identified above, the use of such approaches, at least in isolation, should at a minimum be questioned.

Valuation 6.5 'Short-cut'DCF valuation

Assumptions/facts

Rent	£50,000		
RR yield	7%		
Reversionary yield	7%		
Years to reversion	3		
OMRV	£50,000		
Equated yield	10.00%		
Void period (yrs)	2		
Implied annual growth rate	3.4209%		

Calculation

Term		£50,000	
YP 3 yrs @ 10.00%		2.487	£124,343
Reversion		£55,309	
YP perp. @ 7.00%	14.286		
Def. 5 yrs @ 10.00%	0.621	8.8703	£490,606
			£614,949
Less costs		5%	£29,283.28
			£585,665.58

Contemporary approaches

'Short-cut' DCF

The 'short-cut' DCF has the advantage over the traditional methods of valuation by being more explicit about the expected total return and the implied growth rate used. It also more accurately reflects the weight to be applied to each part of the income flow. How, then, does it perform when dealing with short leases?

Given the information supplied on p. 62 we assume the equated yield for rack-rented property of this class to be 10% (ie 2% above the long-term gilt rate). Given the rack-rented ARY of 7% and the standard rent review interval of five years, this equates to an implied annual growth rate of 3.4209%. Assuming a void period of two years after the end of the lease the valuation can be laid out as in Valuation 6.5.

This is of course rather higher than the valuations that have been produced using the traditional methods. What is the problem? One

Valuation 6.6 Modified 'short-cut'DCF valuation

Assumptions/facts

Rent	£50,000
RR yield	7%
Reversionary yield	10%
Reversion	3
OMRV	£50,000
Equated yield	10.00%
Void period (yrs)	2
Implied annual growth rate	3.4209%

Calculation

Term		£50,000	
YP 3 yrs @ 10.00%		2.487	£124,343
Reversion		£55,309	
YP perp. @ 10.00%	10.00		
Def. 5 yrs @ 10.00%	0.621	6.209	£343,424
			£467,767
Less costs		5%	£22,274.62
			£445,492.36

possible problem is the assumption that has been made about the re-letting of the property in year 5. As with a 'normal' 'short-cut' DCF valuation of a reversionary investment, the assumption made is that the property becomes like a rack-rented property of the same class as the comparable on reversion. This may or may not be the case. If the property is re-let on a short lease the required yield should be rather higher to reflect the lack of security. This alternative is presented as Valuation 6.6.

Although this approach produces an acceptable valuation in terms of the reported figure there is a lot that is questionable about the approach. The two halves of the valuation contain different assumptions. In the term part there is growth assumed at 3.4209% pa; however, the effective income growth rate in the reversion is zero.

What are the alternatives to adjust the valuation? Two are possible. One is to use a lower implied growth rate that implicitly allows for the interruption of income. This would arise naturally if a higher ARY was used in the implied annual growth rate

---

Valuation 6.7 'Short-cut' DCF analysis (or valuation) assuming the Valuation 6.1 figure is correct

Assumptions/facts

Rent	£50,000
RR yield	7%
Reversionary yield	7%
Reversion	3
OMRV	£50,000
Equated yield	25.19%
Void period (yrs)	0
Implied annual growth rate	3.4209%

Calculation

Term		£50,000	
YP 3 yrs @ 25.19%		1.946435581	£97,322
Reversion		£55,309	
YP perp. @ 7.00%	14.28571429		
Def. 3 yrs @ 25.19%	0.509637803	7.280540046	£402,678
			£500,000
Less costs		5%	£23,809.52
			£476,190.48

---

calculation. The other, the preferred method, is to use a different equated yield. The problem with this investment is the inherent additional risk. It is not logical to apply the expected return to this investment as if it were an investment with a secure income flow.

How can this alternative yield be arrived at? The answer is it depends on whether there is sufficient comparable evidence available. If there is little available then it is likely that the equated yield might be arrived at subjectively. A valuer might use an equated yield of between 12% and 15%. Alternatively, if a comparable is available the equated yield can be calculated by trial and error. Let us assume that our initial yield valuation in Valuation 6.1 was actually a transaction. When the equated yield is analysed (using the goal seek function of Excel and fixing all the other assumptions) this produces a figure of 25.19%. Note that the void period has been reduced to zero in the analysis. This is because the void allowance in our calculations to date has been an explicit allowance for risk. This can alternatively be included within the

Table 6.3 Sensitivity summary

	Current values	s1	s2	s3	High comb.	Low comb.
Variables:						
Void period	2	1	2	2	1	3
Equated yield	12.00%	12.00%	11.00%	12.00%	11.00%	13.00%
RR ARY yield	7.00%	7.00%	7.00%	8.00%	6.00%	8.00%
Result:						
Valuation	£541,362.32	£592,601.05	£562,940.22	£541,362.32	£612,063.23	£473,876.12
Difference		9.46%	3.99%	0.00%	13.06%	−12.47%

equated yield and indeed it seems appropriate to do so here. This is illustrated in Valuation 6.7.

These valuations illustrate both the strengths and weaknesses of the 'short-cut' DCF approach. On the one hand the valuation done in this explicit way suggests that the market may be underpricing this type of investment. An equated yield of 25% plus seems very high, particularly if the property will re-let quickly. It is explicitly allowing for risk, however. On the other hand, although a relatively simple technique, there are a lot of pitfalls and potential for mistakes for the practising valuer who may not be experienced in using the method. In particular, would a valuer in practice ever use such a high equated yield? It seems unlikely. Admittedly, if the valuer reinstated the assumption about the void period and used a slightly higher equated yield, a final valuation closer to the figure produced will result. The instability of the method in these circumstances tends to suggest that this method would not be used which, given its advantages and openness, is unfortunate.

These observations underline the fact that the 'short-cut' DCF, being derived from the traditional approach, requires good market evidence to be available and to be absolutely reliable.

Given the variety of approaches that can be taken, carrying out a sensitivity analysis was difficult in that it was not easy to determine the 'base' valuation. It was decided to use an adaptation of Valuation 6.5, ie assuming a void period but with an assumed equated yield, this yield being used in the valuation alone, not in the analysis of implied annual growth. It can be seen from Table 6.3 that the approach is reasonably stable, albeit that the valuation is higher than that produced by the conventional approaches.

'Full' DCF

An alternative approach that might be employed where market evidence is limited is the 'full' DCF approach. The advantage here is that explicit assumptions about the future performance of the investment can be incorporated into the appraisal (see Valuation 6.8).

Given the characteristics of the 'full' DCF technique that we have explored in Part 1, we clearly need to be cautious about using this method. As we know, it is vulnerable to different assumptions made by the valuer. Despite this, the sensitivity analysis of the valuation illustrates that, at these values at least, changing the individual variables by a whole percentage point in each case produces relatively little variation in the outcome (see Table 6.4).

Valuation 6.8 'Full'DCF approach

Assumptions/facts

Rent	£50,000	Discount rate	12.00%
RR yield	7%	Forecast annual	
Reversion	3	growth rate	5.0000%
Void	2		
OMRV	£50,000	Terminal yield	10%

Calculation

Year	OMRV £	Lease rent £	Sale value £	CF £	DF	DCF £
0	50,000					
1	52,500	50,000		50,000	0.892857143	44,643
2	55,125	50,000		50,000	0.797193878	39,860
3	57,881	50,000		50,000	0.711780248	35,589
4	60,775	0		0	0.635518078	0
5	63,814	0		0	0.567426856	0
6	67,005	63,814		63,814	0.506631121	32,330
7	70,355	63,814		63,814	0.452349215	28,866
8	73,873	63,814		63,814	0.403883228	25,773
9	77,566	63,814		63,814	0.360610025	23,012
10	81,445	63,814	814,447.31	878,261	0.321973237	282,777
						512,850.10
				Less costs	5%	24,421.43
						488,428.67

When the three main variables, or assumptions, are changed together, however, then the value figure is seen to be more volatile. In fact there are five variables that make a difference to the valuation: rental growth, terminal yield, discount rate, void period and assumption about the holding period (ie the period of explicit analysis of the cash flow).

That this latter factor has an effect on value can be seen from Valuation 6.9. Here the cash flow is terminated at the end of year 5, ie assuming that the property has reached peak value on re-letting and has been sold. With apparently exactly the same assumptions the valuation is reduced by nearly £30,000 or 6%.

Table 6.4 Sensitivity summary of Valuation 6.8

	Current values	s1	s2	s3	High comb.	Low comb.
Variables:						
Discount rate	12.00%	11.00%	12.00%	12.00%	13.00%	11.00%
Growth rate	5.0000%	5.0000%	6.0000%	5.0000%	4.0000%	6.0000%
Terminal yield	10%	10%	10%	9%	11%	9%
Result:						
Valuation	£488,428.67	£522,844.85	£519,292.90	£516,177.90	£411,807.83	£589,846.04
Difference		7.046%	6.319%	5.681%	-15.687%	20.764%

Valuation 6.9 'Full'DCF valuation with shorter cash flow period

Assumptions/facts

Rent	£50,000		Discount rate	12.00%
RR yield	7%		Forecast annual	
Reversion	3		growth rate	5.0000%
OMRV	£50,000		Terminal yield	10%

Calculation

Year	OMRV £	Lease rent £	Sale value £	CF £	DF	DCF £
0	50,000					
1	52,500	50,000		50,000	0.8928571	44,643
2	55,125	50,000		50,000	0.7971939	39,860
3	57,881	50,000		50,000	0.7117802	35,589
4	60,775	0		0	0.6355181	0
5	63,814	0	638,140.78	638,141	0.5674269	362,098
						482,189.78
			Less costs 5%			22,961.42
						459,228.36

The reason why this has occurred is found in the phrase 'apparently the same assumptions'. In fact the assumptions have changed from one valuation to the next. The changes are related to the growth and terminal yield assumptions and their interrelationships. The original valuation assumed that the growth in rents would be 5% pa over the ten years analysed after which the property would be sold at a yield of 10%. The second valuation assumed 5% growth over only the first five years after which the property was sold, again at a yield of 10%. This high yield implies a low, perhaps even zero, long-term growth rate and it is this that has been applied to the second five-year period. This point illustrates just how carefully each explicit assumption must be arrived at. Here what should have been done is that the valuer should have reconsidered the terminal yield used in the shorter explicit cashflow and decided what yield a seven-year-old newly let property should attract instead of a 12-year-old property five years into a lease. Logic suggests, all things being equal, that this yield should be lower.

This approach, although not tied to market transactions, has a number of advantages in that it lets the valuer explicitly explore the characteristics of the property. Where reasoned and reasonable assumptions can be made the valuation produced is likely not to show a higher level of variance than the traditional approaches. It is thus particularly useful where specific market evidence is limited and where, as a minimum, it can act as a useful alternative to the traditional approaches.

Advanced approaches

Simulation

Arisk-explicit simulation approach to the valuation of the subject is the methodology that is most removed from traditional market techniques. It is a pure investment appraisal method. Can it, then, legitimately be used for valuation? The answer to this is probably yes, subject to an understanding of what the technique represents. It should be accepted that although transaction evidence is important, a transaction represents the result of a series of investors, many of whom will be valuers/surveyors, carrying out appraisals. However, the actual transaction itself may not represent a rational result of those appraisals. A simulation approach such as this represents the valuer's best simulation of how this market behaves or how it should behave. If the valuer has good experience of the marketplace in practice it seems likely that the result will be a very good estimate of how the market will actually behave and consequently may predict the exchange price more accurately.

An illustration of how such a valuation is constructed is given below. The first step is to identify the key variables in the valuation and to use @Risk to ascribe the value limits and the probability distribution, as described in Part 1. This is illustrated in Table 6.5. These variables are then used in the simulation, each inputting into the DCF to calculate the value of each iteration. The base value, based on the most common values for each variable, is shown in Valuation 6.10.

Although the same observations about this type of approach made in Part 1 apply here, it is also clear that this approach can produce an acceptable valuation. The valuation is reasonably stable, given that the technique naturally incorporates a sensitivity analysis by way of the standard deviation in its construction.

Table 6.5 The variables used in the @Risk simulation

Variable	@Risk distribution type	Values (min., most likely, max.)	Comments
The projected sale yield at the end of the holding period	Triangular	8%, 10%, 12%	Reflects the range of possible yields at the end of the cash flow. A 10% terminal yield is most likely but if a poor quality tenant with a short lease takes the property then the yield could be higher and vice versa.
The annual rental growth rate (taken as the annual average over the holding period)	Triangular	2%, 5%, 8%	The most likely rental growth rate is 5% but there is assumed to be some lesser probability of higher or lower rates of growth.
The discount rate	Triangular	10%, 12%, 14%	When used as in investment appraisal, this would probably not be a variable. In valuation, however, the wider market needs to be considered. Different investors would have different views on the risk and return balance and thus have different target rates of return.
The void period	Triangular	0 years, 2 years, 4 years	Instead of allowing for the void period in the cash flow, mechanically it is easier to allow for an end deduction from the valuation. This is calculated using the OMRV multiplied by a YP for x years at the discount rate, discounted for three years at the discount rate. Both the capitalisation period and the discount rate are @Risk variables, thus reflecting the risk involved with each. The most likely void period is taken to be two years.

Valuation 6.10 Base 'full' DCF valuation of subject using average values of probability distributions

Assumptions/facts

Rent	£50,000		Discount rate	12.00%
RR yield	7%		Forecast annual	
Reversion	3		growth rate	5.000%
Void	2			
OMRV	£50,000		Terminal yield	10.00%

Calculation

Year	OMRV £	Lease rent £	Sale value £	CF £	DF	DCF £
0	50,000					
1	52,500	50,000		50,000	0.892857143	44,643
2	55,125	50,000		50,000	0.797193878	39,860
3	57,881	50,000		50,000	0.711780248	35,589
4	60,775	57,881		57,881	0.635518078	36,785
5	63,814	57,881		57,881	0.567426856	32,843
6	67,005	57,881		57,881	0.506631121	29,324
7	70,355	57,881		57,881	0.452349215	26,183
8	73,873	57,881	738,727.72	796,609	0.403883228	321,737

	566,963.50
Less void	64,345.58
	502,617.92
Less costs 5%	23,934.19
	£478,683.74

This cash flow represents the valuation produced by the mean values for the variables chosen for the @Risk analysis. Carrying out the simulation using 1,000 iterations produces the following figures:

Mean valuation	£482,738
Standard deviation	£50,554
Coefficient of variance	10.47%

Arbitrage approach

The arbitrage approach can be applied to dealing with short leases. Where there is evidence available to establish an appropriate initial

---

Analysis of comparable using arbitrage price method

Assumptions/facts

Rent	£50,000
RR yield	10%
Reversion	5
OMRV	£50,000
Low risk yield	8.50%
DCY	8.80%

Calculation

Term		£50,000	
YP 5 yrs @ 8.50%		3.941	£197,032
Reversion		£50,000	
YP perp. @ 10.00%	10.000		
Def. 5 yrs @ 8.80%	0.6559	6.559	£327,968
			£525,000
Less costs		5%	£25,000.00
			£500,000.00

---

yield it can offer a number of distinct advantages, particularly over the alternative, traditional approaches:

- It does not require the valuer to make explicit assumptions about the possible void period.
- It does not require the growth rate of the investment to be assumed or forecast.
- It is reasonably insensitive to changes in the key variables used, suggesting that it will be tolerant to errors.

The basic procedure used in the valuation is as described in Part 1. The valuation requires the selection of a low risk yield that will again be derived from the appropriate current loan rate. The yield on reversion used is the same as would be achieved on a rack-rented freehold investment of the same risk characteristics as the subject. In this case this will be a newly let investment with a short-term lease. This type of investment is assumed to sell at a yield of 10%. What is left is to calculate the appropriate discounted capital yield (DCY) for this class of investment. This is done by representing the analysis of a comparable using the arbitrage assumptions made above. This is illustrated in the box above. The DCY is calculated by trial and error, in this case by using the goal seek function of Excel.

Valuation 6.11 Arbitrage valuation of short lease subject property				
Assumptions/facts				
Rent	£50,000			
RR yield	10%			
Reversion	3			
OMRV	£50,000			
Low risk yield	8.50%			
DCY	8.80%			
Calculation				
Term			£50,000	
YP 3 yrs @ 8.50%			2.554	£127,701.12
Reversion			£50,000	
YP perp. @ 10.00%		10.000		
Def. 3 yrs @ 8.80%		0.776	7.765	£388,227.95
				£515,929
Less costs			5%	£24,568.05
				£491,361.02

The valuation of the subject can now be carried out. This is presented as Valuation 6.11.

As noted, the valuations produced are quite stable. This can be illustrated from the sample sensitivity analysis presented in Table 6.6. It is accepted that arbitrarily altering the yield in this way breaks the theoretical integrity of the arbitrage approach but it is an interesting analysis nonetheless. In fact, the approach is far more sensitive in analysis, ie in the assumptions made in examining the comparable evidence.

Summary and conclusions

The full merits and demerits of each approach will be discussed in Chapter 9. At this stage, however, it is useful to review the performance of the models. The valuations produced by the various methods can be reviewed as in Table 6.7.

Table 6.7 is not put forward as being representative of what valuers in the marketplace would or should have produced as valuations of the subject property using each of the approaches analysed. They are the outcomes of the assumptions made by the

Table 6.6 Sensitivity summary

	Current values	s1	s2	s3	High comb.	Low comb.
Variables:						
DCY	8.80%	9.80%	8.80%	8.80%	9.80%	7.80%
Low risk yield	8.50%	8.50%	9.50%	8.50%	9.50%	7.50%
Reversionary yield	10.00%	10.00%	10.00%	11.00%	11.00%	9.00%
Result:						
Valuation	£491,361.02	£481,347.65	£489,212.66	£457,748.21	£446,496.79	£546,193.71
Difference		−2.04%	−0.44%	−6.84%	−9.13%	11.16%

Table 6.7 Performance by valuation approach

Valuation approach	Valuation £	Difference from mean %	s1 %	s2 %	s3 %	High comb. %	Low comb. %
Initial yield	476,190	-2.219	11.11	9.1	n.a.	n.a.	n.a.
Initial yield with void allowance/equivalent yield term and reversion	464,417	-4.637	10.83	6.11	n.a.	20.63	16.62
Split yield term and reversion	464,485	-4.623	4.01	8.12	5.12	17.63	18.48
'Short-cut' DCF (see text)	541,362	11.163	9.46	3.99	0.00	18.06	12.47
'Full' DCF	488,429	0.294	7.05	6.32	5.68	15.69	20.76
Simulation	482,738	-0.875					
Arbitrage	491,361	0.896	2.04	0.44	6.84	9.13	11.16
Mean	**486,997**						

author in the valuations. Moreover, the mean valuation is not meant to indicate the expected market price – in fact it is skewed by the high valuation produced by the 'short-cut' DCF valuation used for the sensitivity analysis.

What Table 6.7 does do is give a feel for the comparative merits of the approaches. It can be seen that the traditional methods tend to produce a lower valuation than the more explicit contemporary and advanced approaches. This perhaps represents the crudeness of the the traditional methods in this context. It also illustrates that, with the exception of the arbitrage approach, the more explicit, less market-based approaches are not much more sensitive to the assumptions made than any of the market approaches. The question arises, how securely can the market-based approaches arrive at values such as their yield assumptions? If these can be derived from the market and if the majority of valuers would arrive at the same analysis then they should be more reliable. If these assumptions are not secured on market facts then there is probably an advantage in being more specific.

Chapter 7

Break Clauses

Problems created for valuers

Break clauses have become an increasingly common feature of commercial leases in the UK. Professor Neil Crosby and his team from the University of Reading have analysed the changing structure of commercial leases for the RICS using the IPD database (RICS, 1998, Paper 2). They found that 15% of the leases created in 1992 that they examined had break clauses, rising to just under 25% of leases created in 1995–96.

Although break clauses add flexibility in the marketplace for tenants, their existence creates problems for valuers. The problems can be summed up in one word: uncertainty – uncertainty as to whether the break will be operated by the tenant, uncertainty as to what will happen if the break is operated. How long will it take to re-let the space? What terms will apply to the new lease? What level of rent-free period and/or incentive will be required? What standard of tenant will be found to take the new space?

The other thing that break clauses do is greatly add to the complexity of the market. While a few years ago valuers only had to deal with differences mainly due to the quality of the accommodation or location, today the market has tended to become far more heterogeneous. Traditional valuation based on the analysis of transactions on a fairly superficial level is most reliable where the market is as homogeneous as possible.

In many respects valuing an income-producing property subject to breaks is similar to valuing a property let on a short lease, save that the degree of uncertainty is perhaps greater. The characteristics of these investments are distinct from short-term investments in that, beyond certain key points in the lease, the investment may return to being similar to a rack-rented freehold with a known covenant. As we have seen from short leases, dealing with uncertainty is not easy, particularly with conventional valuation techniques. We will look at how each of the various families of methods deals with them, starting with the traditional techniques. It will soon become clear that most of the methods have one common flaw: they need to expressly consider that uncertainty to produce a valid outcome.

Because the approaches are relatively similar, many of the observations made in Chapter 6 about the respective qualities of the methods are common to this chapter as well. Only when the findings are distinct are separate calculations shown.

To consider how the various approaches to valuation tackle this type of problem we will consider the following example:

A modern office property let two years ago on a 15-year lease at a rent of £50,000 pa. The lease is on FRI terms, with no rent reviews, but is otherwise standard. The property's current OMRV is now £60,000. The tenant has the option to terminate the lease on the fifth anniversary of its commencement, subject to giving six month's notice, without penalty.

Market evidence:
Rack-rented office investments let on 15-year FRI leases with five-yearly rent reviews to the then OMRV sell at prices equivalent to 7%.

Other information:
- Long-term government securities yield 8%.
- The normal risk premium for this type of investment is 2% above the long-term gilt rate.

Traditional methods

As with most of the traditional methods, if there is sufficient market evidence from investments with similar characteristics then the methods can be successfully applied. If not the, implicit nature of the traditional approach finds it difficult to cope with circumstances that require the valuer to make explicit judgements about the future.

With traditional techniques valuers are restricted to making adjustments to the yield, often without adequate comparable evidence to base their judgements on, or to making deductions from the final valuation to allow for future possible income voids and expenditure. None of this is really satisfactory. There is too great a risk of over- or under-adjusting with this rather crude approach.

Contemporary models

What then are the alternatives? Crosby and his team looked at using a modified 'short-cut' DCF model using a relatively limited range of possible options that are weighted according to their probability of occurrence (RICS, 1998, Paper 2). This limited range of options has the benefit of simplicity but reduces the effectiveness of using this approach. Such an approach can, of course be applied to the traditional models but the explicit characteristics of the 'short-cut' approach lend themselves particularly to this method.

An example of this is illustrated in Valuation 7.1. As can be observed this technique requires a number of different options to be evaluated. In this case three different views of the future are addressed: in the first the break is not operated; in the next the break is operated and there is a re-letting to a good quality tenant after one year; finally, the break is operated and a re-letting takes place after two years to a poorer quality tenant. These valuations are then tabulated and the probability of occurrence of each is assumed (see Table 7.1). These probabilities – arrived at subjectively – are assessed to be a 60% chance of the break not being operated and a 20% chance each for the other two scenarios occurring. What results is a valuation weighted by the assessed risk of the occurrence of the different scenarios.

Although the probability judgement can make a difference to the valuation (this will occur where there is a marked difference between the outcome of the scenarios) in this case the valuation is very stable to changes in these assumptions. This can be seen from Table 7.2 where equal probabilities of occurrence for the three scenarios are applied. This area is one where the valuer should show caution, however.

This approach is a big advance on the traditional, implicit methods in that it does address the key issue of uncertainty. It is a little limited in that only a relatively small number of potential outcomes are explored. The future of this investment may be quite different from the options assessed. This can be corrected by choosing more scenarios, although the problems with assessing the probability of occurrence similarly increase.

Advanced or possible methods

The same sort of approach can be applied to both the arbitrage and 'full' DCF methods. The arbitrage method in particular has similar

---

Valuation 7.1 'Short-cut'DCF

(a) Break clause not operated

Assumptions/facts

Rent	£50,000
RR yield	7%
Reversionary yield	7%
Reversion	3
OMRV	£60,000
Equated yield	10.00%
Implied annual growth rate	3.4209%

Calculation

Term		£50,000	
YP 3 yrs @ 10.00%		2.487	£124,342.60
Reversion		£66,371	
YP perp. @ 7.00%	14.286		
Def. 8 yrs @ 10.00%	0.751	10.733	£712,360.29
			£836,702.89
Less costs		5%	£39,842.99
			£796,859.89

(b) Break clause operated: re-letting to similar tenant after one year

Assumptions/facts

Rent	£50,000
RR yield	7%
Reversionary yield	7%
Reversion	4
OMRV	£60,000
Equated yield	10.00%
Implied annual growth rate	3.4209%

Calculation

Term		£50,000	
YP 3 yrs @ 10.00%		3.170	£158,493.27
Reversion		£68,641	
YP perp. @ 7.00%	14.286		
Def. 4 yrs @ 10.00%	0.683	9.757	£669,753.82
			£828,247.09
Less costs		5%	£39,440.34
			£788,806.75

(c) Break clause operated: property lets to poor tenant after two years

Assumptions/facts

Rent	£50,000
RR yield	7%
Reversionary yield	8%
Reversion	5
OMRV	£60,000
Equated yield	10.00%
Implied annual growth rate	3.4209%

Calculation

Term		£50,000	
YP 3 yrs @ 10.00%		3.791	£189,539.34
Reversion		£70,989	
YP perp. @ 8.00%	12.500		
Def. 5 yrs @ 10.00%	0.621	7.762	£550,983.69
			£740,523.03
Less costs		5%	£35,263.00
			£705,260.03

Table 7.1 Review of valuation scenarios (a), (b) and (c)

Scenario	Valuation £	Probability of occurrence %	Valuation × Probability £
(a) Break clause not operated	796,859.89	60	478,114.40
(b) Break clause operated with re-letting after one year to similar quality tenant	788,806.75	20	157,761.35
(c) Break clause operated with re-letting to poorer covenant after two years	705,260.03	20	141,052.01
Total			£776,927.76

Table 7.2 Review of valuation scenarios (a), (b) and (c) with equal probability of occurrence

Scenario	Valuation £	Probability of occurrence %	Valuation × Probability £
(a) Break clause not operated	796,859.89	33.3333	265,619.70
(b) Break clause operated with re-letting after one year to similar quality tenant	788,806.75	33.3333	262,935.32
(c) Break clause operated with re-letting to poorer covenant after two years	705,260.03	33.3333	235,086.44
Total			£763,641.46

characteristics to both the traditional methods of valuation and the 'short-cut' DCF approach. The possibility of using alternative models drawn from the finance markets has also been considered. These are promising in that they offer a rigorous, rational method of pricing the option, but they also have practical drawbacks in their application in the property market.

An alternative to the above is to go a step beyond the 'short-cut' DCF approach and use a simulation approach tied to a full discounted cash flow, as we have done previously. This is not a new approach. Herd and Lizieri used Monte Carlo simulation to appraise break clauses in a paper presented at an RICS research conference in 1994. There are, however, still many in the industry who would throw their hands up in horror at the thought of using this technique but there are a number of advantages to it. In particular, it provides a way of explicitly addressing the areas of uncertainty in a comprehensive fashion. In addition, it is a transparent technique – the assumptions made by the valuer can be clearly seen.

It is within the capabilities of @Risk to construct a single model which would address the uncertainties created by the existence of break clauses, and particularly those uncertainties related to the key question of whether the break will be operated at all. Such a model would be rather complex and the results difficult to interpret, but things can be simplified by running two simulations and weighting the results, similar to the way the 'short-cut' DCF was conducted. The first simulation, shown in Valuation 7.2, explores the range of possible outcomes that might occur if the break is operated. The second simulation, shown in Valuation 7.3, deals with the assumption that the break is *not* operated. The results of the two @Risk simulations (using the variables given in Table 7.3) are combined in Table 7.4, with a weighting of 60/40 in favour of the assumption that the break will not be operated.

It can be observed that the valuation is somewhat lower than that produced by the probability weighted 'short-cut' DCF. This under-lines the fact that the assumptions made in the two models are different. The simulation approach is more explicit: the valuer has addressed many of the areas of uncertainty about the future and explored the reliability of the assumptions that have been made. Again, this is both the main strength and the main weakness of this method.

Why use this approach? Does it offer much of an advance over, for example, using a 'short-cut' DCF approach with a variety of scenarios? There are a number of objections to using simulation. The technique is relatively complex. It requires valuers to speculate about a number of events that are uncertain. It also requires valuers to assess the probability of a range of possible outcomes occurring. This is perhaps the biggest criticism – in practice these probabilities will have to be determined subjectively.

While these criticisms are valid, there are, however, a number of advantages possessed by the approach that may outweigh them. Essentially, unless good market evidence exists to enable investors' attitudes to be assessed, the very existence of break clauses requires valuers to speculate about the future and to make subjective assess-ments as to its probability of occurrence. Conventional techniques do not give the valuer an adequate framework to address this uncertainty; simulation, on the other hand, does. Constructing a simulation model requires the valuer to explicitly consider what might happen in the future and to come to a reasoned and reasonable opinion of value. It is in providing this decision-making framework that simulation really does offer an advantage over other approaches.

This has probably always been the case. However, in the past the ability to apply this approach was beyond the capabilities of the vast majority of valuers. Technological advances have opened up the possibility of using these techniques in the same way that the

Valuation 7.2 Base assumption for simulation assuming that the break clause is operated

Assumptions/facts

Rent	£50,000	Discount rate	10.00%
RR yield	7%	Forecast annual	
Void allowance	2	growth rate	4.5000%
OMRV	£60,000	Terminal yield	10.00%

Calculation

Year	OMRV £	Lease rent £	Sale value £	CF £	DF	DCF £
0	60,000					
1	62,700	50,000		50,000	0.909090909	45,455
2	65,522	50,000		50,000	0.826446281	41,322
3	68,470	50,000		50,000	0.751314801	37,566
4	71,551	68,470		68,470	0.683013455	46,766
5	74,771	68,470		68,470	0.620921323	42,514
6	78,136	68,470		68,470	0.56447393	38,650
7	81,652	68,470		68,470	0.513158118	35,136
8	85,326	68,470		68,470	0.46650738	31,942
9	89,166	85,326		85,326	0.424097618	36,187
10	93,178	85,326		85,326	0.385543289	32,897
11	97,371	85,326		85,326	0.350493899	29,906
12	101,753	85,326		85,326	0.318630818	27,188
13	106,332	85,326	1,063,317.66	1,148,644	0.28966438	332,721

	778,248.52
Less value of void	89,280.37
	688,968.15
Less costs 5%	122,088.38
	£656,160.14

Mean	£661,007.70
SD	£65,768.18

Valuation 7.3 Base assumption for simulation assuming that the break clause is not used

Assumptions/facts

Rent	£50,000	Discount rate	10.00%
RR yield	7%	Forecast annual	
Reversion	3	growth rate	4.5000%
OMRV	£60,000	Terminal yield	10.00%

Calculation

Year	OMRV £	Lease rent £	Sale value £	CF £	DF	DCF £
0	60,000					
1	62,700	50,000		50,000	0.909090909	45,455
2	65,522	50,000		50,000	0.826446281	41,322
3	68,470	50,000		50,000	0.751314801	37,566
4	71,551	68,470		68,470	0.683013455	46,766
5	74,771	68,470		68,470	0.620921323	42,514
6	78,136	68,470		68,470	0.56447393	38,650
7	81,652	68,470		68,470	0.513158118	35,136
8	85,326	68,470		68,470	0.46650738	31,942
9	89,166	85,326		85,326	0.424097618	36,187
10	93,178	85,326		85,326	0.385543289	32,897
11	97,371	85,326		85,326	0.350493899	29,906
12	101,753	85,326		85,326	0.318630818	27,188
13	106,332	85,326	1,063,317.66	1,148,644	0.28966438	332,721
						778,248.52
			Less costs	5%		37,059.45
						£741,189.07

	Mean	£744,362.40
	SD	£55,991.84

Table 7.3 The variables used in the @Risk simulation

Variable	@Risk distribution type	Values (min., most likely, max.)	Comments
The projected sale yield at the end of the holding period	Triangular	9%, 10%, 11%	Reflects the range of possible yields at the end of the cash flow. A 10% terminal yield is most likely but if a poor quality tenant with a short lease takes the property then the yield could be higher and vice versa.
The annual rental growth rate (taken as the annual average over the holding period)	Triangular	2%, 4.5%, 7%	The most likely rental growth rate is 4.5% but there is assumed to be some lesser probability of higher or lower rates of growth.
The discount rate	Triangular	10%, 12%, 14%	When used as in investment appraisal, this would probably not be a variable. In valuation, however, the wider market needs to be considered. Different investors would have different views on the risk and return balance and thus have different target rates of return.
The void period (applies to the break clause operated simulation only)	Triangular	0 years, 2 years, 4 years	Instead of allowing for the void period in the cash flow, mechanically it is easier to allow for an end deduction from the valuation. This is calculated using the OMRV multiplied by a YP for x years at the ARY, discounted for three years at the discount rate. Both the capitalisation period and the discount rate are @Risk variables, thus reflecting the risk involved with each. The most likely void period is taken to be two years.

Table 7.4 Production of final valuation figure using probability assessment for each scenario

Cash flow assumptions	Mean valuation (from @Risk simulation) £	Probability %	Valuation × Probability £
Break clause not operated	744,362.40	60	446,617.44
Break clause operated	661,007.70	40	264,403.08
Total (valuation figure)			**£711,020.52**

availability of valuation and appraisal software have made the physical calculation of values so much easier. It is surely a logical thing to use the technology available to improve the quality of the output of the valuation process.

Summary and conclusions

Again the main summary will be left until Chapter 9. Briefly, however, the observations made about the valuation of short leases and the relative merits of the various approaches apply to the valuation of break clauses. Essentially an explicit approach is required, and to be successful this explicit approach must extend to the question of risk so some kind of scenario approach seems inevitable. This in turn takes valuers further away from direct comparison with market evidence as the basis for their value opinion, though fundamentally the market will be the final arbiter of the validity of the value figure.

These two areas (ie short leases and break clauses) are where the option pricing models have the greatest potential for market application. There are real problems, however, in, say, reliably using the principles of arbitrage price theory or risk-neutral approaches in that both derive their pricing ability by comparing the option with assets of known price and equal cash flow and risk. Although theoretically this can be done, with the diversity of possible outcomes in these situations it is difficult to be sure that the comparable is truly of similar worth to the subject. In this regard the multi-path models, such as Monte Carlo simulation, must be considered superior for the production of a rational appraisal of value.

Chapter 8

Over-Rented Investments

Problems posed for valuers

The genesis of over-rented property investments in the early 1990s was the catalyst for the re-examination of many of the traditional valuation methods used for investment property. Over-rented investments reached their peak in the mid-1990s and have declined in importance in terms of the numbers seen by commercial valuers. Notwithstanding this, it is important to look at over-rented investments for two main reasons. First, the UK commercial market structure, and particularly the lease structures, will allow over-rented investments to recur. Second, over-rented investments expose many of the fundamental weaknesses in the traditional approaches.

Persistently over-rented investments require two principal characteristics – other than the basic fall in rental values! They require long leases, preventing the renewal of the lease allowing the adjustment of the market rent. They also require upwards-only rent review clauses, or 'ratchet' clauses, that stop the rents slipping back to the OMRV. Both of these are peculiar to the UK investment market meaning that, although most of the rest of the market economies of the world suffered similar problems, it was only in the UK that over-rented investments presented particular difficulties for valuers. It has already been noted that our traditional valuation techniques have evolved to deal with the peculiarities of our investment market. It should, therefore, be of little surprise to find that it was UK techniques that found themselves most exposed and affected by the change in investment characteristics.

The main problem that the traditional valuation models have with over-rented investments is related to the overloading of assumptions within the all-risks yield, in particular those concerned with the long-term rental growth assumptions. The ARY allows the valuer to make assumptions about the future implicitly; over-rented investments really require explicit assumptions to be made.

The best way to illustrate the problem is to examine an over-rented investment. The following details will be used throughout this chapter as the base example:

(a)

OMRV (R$_r$) £50,000
Lease rent (R$_l$) £40,000

Term

Reversion

∞

Long-term rental
growth line

(b)

Lease rent (R$_l$) £50,000
OMRV (R$_r$) £40,000

Term

Reversion

∞

Long-term rental
growth line

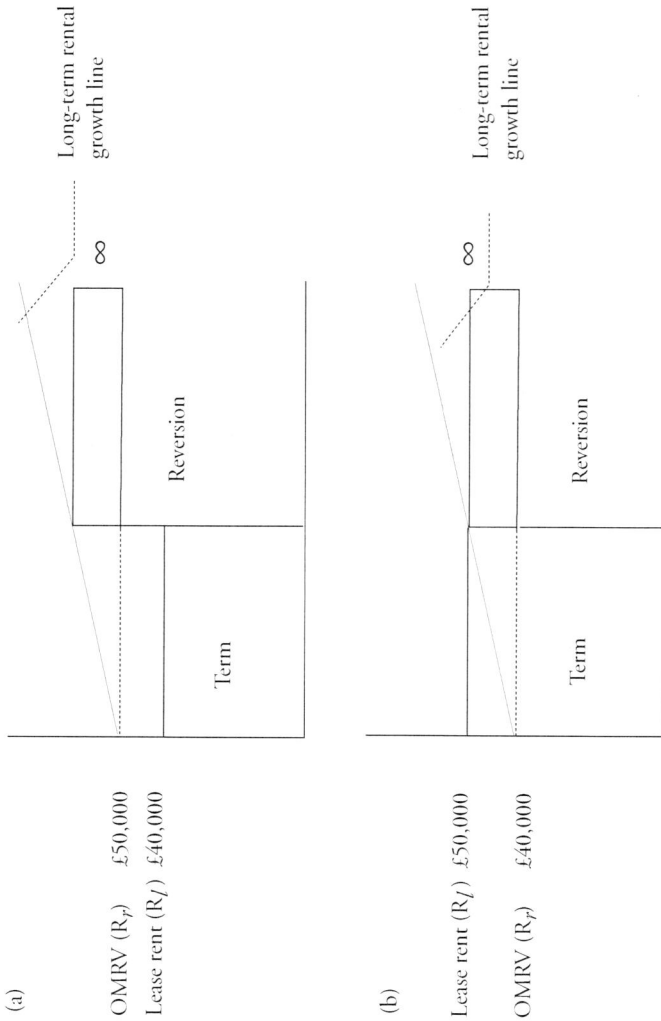

Figure 8.1 Income characteristics of (a) a reversionary, and (b) an over-rented investment

A modern office property let two years ago on a 15-year lease at a rent of £50,000 pa. The lease is on FRI terms, with upwards-only rent reviews to the current OMRV every five years. All other terms are typical for this type of property. The property's current OMRV is £40,000.

Market evidence:
Rack-rented office investments let on 15-year FRI leases with five-yearly rent reviews to the then OMRV sell at prices equivalent to 7%.

Other information:
• Long-term government securities yield 8%.
• The normal risk premium for this type of investment is 2% above the long-term gilt rate.

There was a lot of talk, in the early days of over-rented investments, that the problems of valuation were related to calculating the relative risk of this kind of investment. This is partly true but in essence the main valuation problems were related to the inherent inaccuracies and inconsistencies of the traditional valuation techniques.

Figure 8.1 provides a diagram of the income characteristics of (a) a reversionary and (b) an over-rented investment. The latter type of investment is quite different from both a rack-rented and a reversionary investment. Although both, in practice, are composed of a fixed income component and a growth component, they behave in slightly different ways. With a reversionary investment the yield used in the valuation, where the yield is below the equated yield or expected total return, assumes growth from the current OMRV. This assumes that the term rent is below the OMRV and that adjustment will be made to the new OMRV on reversion (although this is done implicitly). The yields used in traditional valuation overvalue the term by assuming growth in the contractual rent, but balance out by undervaluing the reversion. The technique is a bit of a 'fudge' but by trial, error, practice and experience it actually works.

With the over-rented investment, again it is the OMRV that grows but it is only when the OMRV rises above the contractual rent that the actual income flow will change. There is likely to be a longer period where the investment behaves like a fixed interest security. In addition the overall rate of growth in the investment

Table 8.1 Methodologies employed by the valuers in a 1996 valuation of an over-rented office investment

Methodology employed	Number of occurrences
Traditional (growth implicit) top slice	5
Traditional (growth implicit) top slice with void and reversion after main lease ends	2
Traditional (growth implicit) term and reversion – different yield for term and reversion	2
Initial yield	1
Growth explicit 'short-cut' DCF	1
Total	**11**

Source: Havard (1999a)

will be greatly reduced when compared with a similar rack-rented or reversionary investment. Essentially where the traditional valuation techniques fall down is that it is much harder to judge the 'fudge'. Traditional techniques, particularly where the equivalent yield is used, are a balancing act: both sides of the value calculation are wrong but the final answer is 'correct'. With an over-rented investment, particularly with poor market evidence, it is all too easy to get all three components incorrect!

The traditional valuation approaches were adapted and developed to deal with these types of investment. The basic aim of this was to make them more reliable and less prone to error. To an extent they succeeded but their fundamental inadequacies were still exposed.

There is also apparently a lack of agreement as to which is the best method to use. In research carried out by the author in 1996 an over-rented office property was valued individually by a group of 11 experienced commercial valuers. Not only did they produce quite a wide range of valuations, they also used a number of different valuation approaches (see Table 8.1). It should be noted that, although the participants received common data and viewed the same market evidence, they often used quite different yields even if they were using the same basic approach!

Traditional approaches

There are three basic traditional approaches employed to value over-rented investments:

- initial yield approach;
- 'top slice' approach;
- term and reversion approach.

The merits and demerits of each will be considered in turn.

Initial yield approach

As in the general observations made in Part 1, the basic initial yield approach has much to commend it in terms of its simplicity and directness. However, the problem with using the initial yield approach to value over-rented investments is finding transaction evidence that is truly comparable. Although over-rented investments are sold in the open market, the characteristics of each tend to be individual. The underlying commercial property market is highly heterogeneous; the addition of the complexity of the characteristics of over-rented investments greatly magnifies these differences. Each individual investment has its own special features: the degree of over-rent, the level of security offered by the lease covenant, the length of time that the property will stay over-rented. This possibility of variance reduces the chances of being able to securely decide on the correct ARY for the subject property by analysing transaction evidence. This factor exacerbates the other characteristic of initial yield valuations, namely that the assumptions made by the valuer are implicit. With an initial yield valuation it is not obvious what the valuer has or has not assumed about the property and thus a third party to the valuation would find it hard to assess its validity.

A valuation using the initial yield approach is presented as Valuation 8.1. The yield of 9% used is assumed to have been determined from market evidence.

Using the same limited sensitivity analysis as previously illustrates that the method is reasonably sensitive to the yields adopted, as might be expected (see Table 8.2). This sensitivity has a natural limit due to the relatively high yields used for this type of analysis. As the yield rises so the effect of changing individual variables decreases.

```
Valuation 8.1  Initial yield approach

Assumptions/facts
Rent                      £50,000
RR yield                      7%
Reversion                       3
OMRV                      £40,000
Initial yield               9.00%

Calculation
Reversion                                £50,000
YP perp. @ 9.00%                          11.111      £555,555.56
Less costs                                    5%      £26,455.03
                                                      £529,100.53
```

Table 8.2 Sensitivity summary

	Current values	s1	s2
ARY	9.00%	8.00%	10.00%
Result:			
Valuation	£529,100.53	£595,238.10	£476,190.48
Difference		12.50%	−10.00%

In terms of mathematical correctness, the method has short-comings. The fact that the OMRV is ignored plus the application of a yield below the equated yield implies that the valuer is assuming growth in the term rent from the start, albeit at a slow rate. The implication of growth may be correct over the long term, ie over the life of the investment as a whole, but is incorrect in terms of the true characteristics of the investment that will see zero growth in the lease rent until the OMRV rises sufficiently. This illustrates the degree of 'fudge' that this method represents. As long as the market evidence is good enough, the method can produce a satisfactory valuation, but one that is uninformative and that does not represent the true characteristics of the investment. It was largely the former problem that caused valuers in the early 1990s to develop other approaches to deal with the problem.

'Top slice' approach

This method is generally accepted as that which is the most popularly used by commercial valuers. It combines the capitalisation of the OMRV at relatively low yield – quite commonly this is the rack-rented ARY – into perpetuity with the addition of the overage (the rent received above the OMRV) capitalised for a limited period at a higher yield. This approach is ostensibly to account for the different risk that each part of the income represents. In reality it represents the need to adjust for the different growth characteristics in each part of the income flow.

There are a number of issues and questions raised in this method. First there is the question of what yield to apply to each part of the income flow. Second, there is the question of how long to capitalise the overage for. In the original application of this method, the overage was commonly capitalised from the date of valuation for the complete unexpired term of the lease. It was then realised that the valuations being produced were too high. It was argued that the use of the rack-rented ARY for the bottom slice income was leading to a double counting of the value of the top slice due to the implicit growth assumptions contained within the ARY. This led to the capitalisation of the overage for only a limited period, usually to the first or second rent review, the latter when the level of rental growth was too low or the overage was too large for the OMRV to overtake the lease rent by the first review. This actually produces the inconsistent result that the worse the market the higher the valuation! In fact, the main reason for limiting the capitalisation of the overage was simply that the traditional models used do not reflect the true characteristics of the investment, and thus have to be massaged in order to produce a more acceptable valuation.

A 'top sliced' valuation is illustrated in Valuation 8.2. The OMRV has been capitalised at the rack-rented ARY. The top slice income, the difference between the passing rent and the OMRV, is capitalised for eight years, representing the second rent review following the valuation. The yield used at 10% represents a 'non-growth' yield, ie it is similar to the target rate of return that an investor would seek for a building of this quality.

Despite the shortcomings of the method identified above, the technique is reasonably successful as a method of valuation. It is reasonably stable to individual changes in the key variables, other than the OMRV capitalisation rate (see Table 8.3). In combination, however, the approach is quite sensitive to adjustments in the variables used.

Table 8.3 Sensitivity summary

	Current values	s1	s2	s3	High comb.	Low comb.
Variables:						
Top slice yield	10.00%	11.00%	10.00%	10.00%	9.00%	11.00%
OMRV yield	7.00%	7.00%	8.00%	7.00%	6.00%	8.00%
Capitalisation period	8.00	8.00	8.00	3.00	13.00	3.00
Result:						
Valuation	£595,026.51	£593,228.38	£526,999.30	£567,901.99	£706,224.48	£499,463.95
Difference		−0.30%	−11.43%	−4.56%	18.69%	−16.06%

Valuation 8.2 Traditional 'top slicing'approach

Assumptions/facts

Rent	£50,000
RR yield	7%
OMRV	£40,000
Bottom slice yield	7.00%
Top slice yield	10.00%
Period of capitalisation of top slice (yrs)	8

Calculation

Bottom slice		£40,000	
YP perp. @ 7.00%		14.286	£571,429
Top slice		£10,000	
YP 8 years @ 10.00%		5.335	£53,349.26
			£624,778.00
Less costs		5%	£29,751.33
			£595,026.51

Term and reversion approach

The third alternative approach frequently used by valuers is the term and reversion method, using either a split yield or, less frequently, an equivalent yield approach.

Equivalent yield

The principal of the approach is the same as with the equivalent yield approach in a term and reversion valuation. A single, adjusted, all-risks yield is used to capitalise both the term and the reversionary elements of the investment. The reversion is the current OMRV. The period to reversion depends upon the judgement of the valuer. Commonly it is until the next rent review but the valuer may choose to delay this until the second or later review. Generally the longer the period, the higher the value will be.

Again this method is a fudge, though it reflects the actual income pattern better than the 'short-cut' approach. It is very hard to judge the correct equivalent yield to apply, though it is possible to derive

Table 8.4 Sensitivity summary

	Current values	s1	s2	High comb.	Low comb.
Variables:					
Capitalisation period	8	3	8	13	3
Equivalent yield	8.00%	8.00%	9.00%	7.00%	9.00%
Result:					
Valuation	£530,920.37	£500,734.26	£475,992.99	£623,814.36	£447,387.99
Difference		−5.69%	−10.35%	17.50%	−15.73%

Valuation 8.3 Equivalent yield valuation			
Assumptions/facts			
Rent	£50,000		
RR yield	7%		
Reversion	8		
OMRV	£40,000		
Equivalent yield	8.00%		
Calculation			
Term		£50,000	
YP 8 yrs @ 8.00%		5.747	£287,331.95
Reversion		£40,000	
YP perp. @ 8.00%	12.500		
Def. 8 yrs @ 8.00%	0.540	6.753	£270,134.44
			£557,466.39
Less costs		5%	£26,546.02
			£530,920.37

this from market evidence. The valuation produced is relatively sensitive to the yield choice.

An example of this approach is presented in Valuation 8.3 and Table 8.4. Here it is assumed that the yield has been chosen subjectively. If the same valuation was to be produced as the short-cut approach detailed in Valuation 8.1, a yield of 6.66% would be required (this was found by trial and error using the goal seek function of Excel). In the absence of market evidence it seems unlikely that a valuer would have subjectively chosen a yield this low. It is likely that valuers who choose this method will have a tendency to undervalue in the absence of contrary evidence.

Split yield term and reversion

The alternative approach is to use a split yield term and reversion model. This has three advantages. First, it is less sensitive to yield choice. Second, it correctly reacts to market conditions in that the lower the expected growth, the lower the value (by delaying the reversion to OMRV at the rack-rented ARY, the highest value being in the reversion). Finally, a yield which reflects the true characteristics of the two parts of the investment can be used, namely a high,

Table 8.5 Sensitivity summary

	Current values	s1	s2	s3	High comb.	Low comb.
Variables:						
Capitalisation period	8	3	8	8	13	3
Term yield	10.00%	10.00%	9.00%	10.00%	9.00%	12.00%
Reversionary yield	7.00%	7.00%	7.00%	8.00%	7.00%	8.00%
Result:						
Valuation	£506,237.49	£524,581.48	£535,044.67	£474,503.09	£532,963.82	£461,658.90
Difference		3.62%	5.69%	−6.27%	5.28%	−8.81%

Valuation 8.4 Split yield term and reversion approach

Assumptions/facts

Rent	£50,000
RR yield	7%
Reversion	8
OMRV	£40,000
Term yield	10.00%
Reversionary yield	7%

Calculation

Term		£50,000	
YP 8 yrs @ 10.00%		5.335	£266,746.31
Reversion		£40,000	
YP perp. @ 7.00%	14.286		
Def. 8 yrs @ 10.00%	0.463	6.620	£264,803.05
			£531,549.36
Less costs		5%	£25,311.87
			£506,237.49

---

zero-growth yield for the term, and the conventional growth-implicit yield for the reversion.

The simple fault with the method is that it produces too low a valuation when compared with the other methods. Even using a three-year period for the term, the valuation is still some £70,000 below the top slice approach. To correct this, either the reversionary or the term yield would have to be reduced. If this is done, the superficial logic that underlies the approach breaks down.

Due to these weaknesses it is not advisable to use either of these approaches.

Contemporary approaches

'Short-cut' DCF

The 'short-cut' DCF has a number of advantages over the methods considered up to this point:

- It is technically superior because it explicitly considers when the OMRV will exceed the lease rent.

Table 8.6 Sensitivity summary

	Current values	s1	s2	s3	High comb.	Low comb.
Variables:						
Term yield	10.00%	9.00%	10.00%	10.00%	9.00%	11.00%
OMRV yield	7.00%	7.00%	8.00%	7.00%	6.00%	8.00%
Period until OMRV>LR	8.00	8.00	8.00	3.00	3.00	13.00
Result:						
Valuation	£586,319.11	£621,022.58	£544,784.73	£570,714.18	£662,869.41	£511,307.13
Difference		5.92%	−7.08%	−2.66%	13.06%	−12.79%

Valuation 8.5 'Short-cut'DCF valuation of the subject

Assumptions/facts

Rent	£50,000		
RR yield	7%		
Reversionary yield	7%		
Reversion	8		
OMRV	£40,000		
Equated yield	10.00%		
Implied annual growth rate	3.4209%		

Calculation

Term		£50,000	
YP 3 yrs @ 10.00%		5.335	£266,746.31
Reversion		£52,351	
YP perp. @ 7.00%	14.2857		
Def. 8 yrs @ 10.00%	0.4665	6.664	£348,887.91
			£615,634.22
Less costs		5%	£29,315.92
			£586,318.31

- It also correctly treats the term income as being 'no growth'.
- It does not require transaction evidence from over-rented investments – though there are question raised about the yields used.

The basic approach to the valuation is as described in the preceding sections. The rack-rented ARY on similar properties must be found, the EY assumed and the implied annual growth rate calculated. As before the ARY is 7%, the EY is 10% and the growth rate is thus 3.4209%. The growth rate is then applied to the OMRV to calculate when the rent will rise to exceed the lease rent. In this case this will not occur until the second rent review.

The valuation is as presented in Valuation 8.5.

Technically, the valuation should not change if the variables within the model alter. If the equated yield rises then so does the growth rate to compensate. The model is sensitive to changes in the ARY chosen, however, as this will alter the growth rate and would delay the period at which the rent would rise. It is possible to alter these variables when applied to the subject, ie using a different equated yield with the growth rate as derived from the rack-rented

Table 8.7 Sensitivity summary

	Current values	s1	s2	s3	High comb.	Low comb.
Variables:						
Terminal yield	9.00%	8.00%	9.00%	9.00%	8.00%	10.00%
Growth rate	5.00%	5.00%	4.00%	5.00%	6.00%	4.00%
Discount rate	10.00%	10.00%	10.00%	11.00%	9.00%	11.00%
Result:						
Valuation	£545,787.54	£582,255.47	£524,286.46	£516,421.53	£644,466.91	£471,285.23
Difference		6.68%	−3.94%	−5.38%	18.08%	−13.65%

Valuation 8.6 'Full'DCF valuation of over-rented investment

Assumptions/facts

Rent	£50,000	Discount rate	10.00%
RR yield	7%	Forecast annual	
Reversion	3	growth rate	5.0000%
OMRV	£40,000	Terminal yield	9%

Calculation

Year	OMRV £	Lease rent £	Sale value £	CF £	DF	DCF £
0	40,000					
1	42,000	50,000		50,000	0.9090909	45,455
2	44,100	50,000		50,000	0.8264463	41,322
3	46,305	50,000		50,000	0.7513148	37,566
4	48,620	50,000		50,000	0.6830135	34,151
5	51,051	50,000		50,000	0.6209213	31,046
6	53,604	50,000		50,000	0.5644739	28,224
7	56,284	50,000		50,000	0.5131581	25,658
8	59,098	50,000	656,646.86	706,647	0.4665074	329,656
						573,076.92
			Less costs	5%		27,289.38
						£545,787.54

ARY, though strictly the ARY should stay the same. It is thus instructive to carry out a sensitivity analysis (see Table 8.6). Freezing the growth rate illustrates that the valuation is relatively insensitive to miscalculation of the key variables.

'Full' DCF

An alternative approach, particularly where the market evidence is particularly thin, is to use a fully explicit DCF approach. The approach offers many of the advantages and disadvantages identified in earlier sections. The advantages are related to the ability to make explicit assumptions about the future growth rates and other characteristics of the investment. The disadvantages relate to the individual nature of these explicit assumptions and the difficulties in forecasting key factors such as rental growth.

Table 8.8 The variables used in the @Risk simulation

Variable	@Risk distribution type	Values (min., most likely, max.)	Comments
The projected sale yield at the end of the holding period	Triangular	9%, 10%, 11%	Reflects the range of possible yields at the end of the cash flow. A 10% terminal yield is most likely but if a poor quality tenant with a short lease takes the property then the yield could be higher and vice versa. This yield is higher than used in the base 'full' DCF because the analysis period is longer (see below). The property being sold at the end of the cash flow will be older and thus less competitive.
The annual rental growth rate (taken as the annual average over the holding period)	Triangular	2%, 4.5%, 7%	The most likely rental growth rate is 4.5% but there is assumed to be some lesser probability of higher or lower rates of growth. The growth rate used is lower than that used in the 'full' DCF analysis which forms the basis of this valuation. The reason for this is that the analysis period here is longer. In the later years the growth rate will be lower relative to the market as the property ages. The overall growth rate is reduced to explicitly compensate for this.

| The discount rate | Triangular | 9%, 10%, 11% | When used in investment appraisal, this would probably not be a variable. In valuation, however, the wider market needs to be considered. Different investors would have different views on the risk and return balance and thus have different target rates of return. |
| The analysis period | n.a. | Whole lease | The reason for extending the period of analysis to the full extent of the lease is to explore the effect of lower rates of growth on the value. At the lowest rates of growth the OMRV will not exceed the lease rent until the end of the lease. Extending the analysis period allows this scenario to be explored. This extension has implications for the growth rate used and for the terminal yield (see above). |

Making what appear to be 'reasonable' assumptions, the valuation of our subject property using a 'full' DCF approach is presented in Valuation 8.6. (The accompanying sensitivity analysis is presented in Table 8.7.)

Advanced approaches

Both the simulation and the arbitrage approach can be applied to the valuation of over-rented investments, through both display slightly different qualities. The main advantages of each, respectively, are that in simulation the uncertainty about the future performance of the investment can be addressed, whereas with arbitrage the valuation produced is very stable, showing high levels of consistency when the variables are changed.

Simulation approach

One of the key factors in assessing the value of an over-rented investment is assessing the point at which the investment moves from behaving like a fixed income security to becoming a growth investment, ie the point where the OMRV grows sufficiently to exceed the lease rent. This obviously depends upon the degree of over-rent and also on the rate of rental growth. The traditional approaches, being implicit in their assumptions, have difficulties in addressing this. The growth-explicit approaches are better but still have flaws. The 'short-cut' DCF, for all its qualities, still uses an implied long-term growth rate, not a forecast of the actual expected growth rate. In practice, the 'short-cut' approach may not correctly predict the point at which the changeover takes place. If it is required that this be explicitly addressed then a more explicit forecast approach, such as a 'full' DCF, may be appropriate. As we have seen, however, the 'full' DCF approach relies on a single-point forecast. There is no indication of the reliability of the assumption, nor any testing of the effect of variations in the forecast. This is something that simulation, of course, allows you to do.

The assumptions used in preparing the simulation are laid out in Table 8.8. The comments related to the period of analysis in particular are to be noted.

The simulation thus allows some of the key issues to be explored explicitly, far more so in fact than with any of the other methods. Although this does expose the valuer to a high degree, this method allows the exploration of many of the possible outcomes that the

---

Valuation 8.7 Simulation-based valuation of the subject property

Assumptions/facts

Rent	£50,000		Discount rate	10.00%
RR yield	7%		Forecast annual	
Reversion	3		growth rate	4.50%
OMRV	£40,000		Terminal yield	10.00%

Calculation

Year	OMRV £	Lease rent £	Sale value £	CF £	DF	DCF £
0	40,000					
1	41,800	50,000		50,000	0.909090909	45,455
2	43,681	50,000		50,000	0.826446281	41,322
3	45,647	50,000		50,000	0.751314801	37,566
4	47,701	50,000		50,000	0.683013455	34,151
5	49,847	50,000		50,000	0.620921323	31,046
6	52,090	50,000		50,000	0.56447393	28,224
7	54,434	50,000		50,000	0.513158118	25,658
8	56,884	50,000		50,000	0.46650738	23,325
9	59,444	56,884		56,884	0.424097618	24,124
10	62,119	56,884		56,884	0.385543289	21,931
11	64,914	56,884		56,884	0.350493899	19,938
12	67,835	56,884		56,884	0.318630818	18,125
13	70,888	56,884	708,878.44	765,762	0.28966438	221,814
						572,678.56
			Less costs 5%			27,270.41
						£545,408.15
			Mean			£548,259
			Standard deviation			£32,587
			Coefficient of variance			5.944%

future holds. It makes the valuer address these issues and come up with reasonable solutions. The valuation is also remarkably stable. In the sensitivity analysis that was carried out, the probability distribution of the variables was shifted by 1 percentage point (an example of this is given in Table 8.9). The variables were tested individually and in combination, as previously. The sensitivity

Table 8.9 Example of assumptions made within high valuation combination in the sensitivity analysis

Variable	@Risk distribution type (where appropriate)	Values (min., most likely, max.)
The projected sale yield at the end of the holding period	Triangular	8%, 9%, 10%
The annual rental growth rate (taken as the annual average over the holding period)	Triangular	3%, 5.5%, 8%
The discount rate	Triangular	8%, 9%, 10%
The analysis period	n.a.	Whole lease

summary in Table 8.10 illustrates just how stable the simulation is. It reinforces the attractions of using this approach where market evidence is limited or contradictory or where a check on the valuation is required.

The outcome of the simulation is presented in Valuation 8.7.

Arbitrage approach to valuation

The arbitrage valuation of the subject displays the qualities that we have come to expect from the method. It is stable and relatively easy to calculate. Again it works best where there is good market evidence but can be used to produce a stable valuation where direct comparables are not available. This is illustrated in Valuation 8.8. Here the yields derived from a previous example have been used to value the subject property. There has been an adjustment to allow for the fact that the OMRV is not expected to exceed the lease rent until the second rent review. This is the only concession to the need to explicitly address the future. As can be seen from the sensitivity analysis (see Table 8.11), this adjustment, does, in fact, make very little difference to the overall valuation.

Despite the stability of the figures, there are some questions about the valuation produced. For one thing, it is rather low. Secondly, it suffers from similar problems to the traditional approaches in that it does not reveal what assumptions are contained within it. This can be addressed by converting the arbitrage approach to a growth-

Table 8.10 Sensitivity summary

	Current values	s1	s2	s3	High comb.	Low comb.
Variables:						
Discount rate	10%	9.00%	10.00%	10.00%	9.00%	11.00%
Growth rate	4.50%	4.50%	5.50%	4.50%	5.50%	3.50%
Terminal yield	10.00%	10.00%	10.00%	9.00%	9.00%	11.00%
Result:						
Valuation	£548,259.00	£547,536.20	£581,176.30	£569,521.40	£605,824.70	£502,425.20
Difference		-0.13%	6.00%	3.88%	10.50%	-8.36%

Table 8.11 Sensitivity summary

	Current values	s1	s2	s3	High comb.	Low comb.
Variables:						
CY	9.15%	8.15%	9.15%	9.15%	8.15%	10.15%
Low risk yield	8.50%	8.50%	7.50%	8.50%	7.50%	9.50%
Years of overage	8	8	8	3	13	3
Result:						
Valuation	£538,668.69	£559,309.80	£549,055.38	£540,125.87	£583,474.32	£526,682.42
Difference		3.83%	1.93%	0.27%	8.32%	−2.23%

Valuation 8.8 Arbitrage valuation of the subject			
Assumptions/facts			
Rent	£50,000		
RR yield	7%		
Reversion	8		
OMRV	£40,000		
Low risk yield	8.50%		
DCY	9.15%		
Calculation			
Term		£50,000	
YP 8 yrs @ 8.50%		5.639	£281,959
Reversion		£40,000	
YP perp. @ 7.00%	14.286		
Def. 8 yrs @ 9.15%	0.4964	7.091	£283,643
			£565,602
Less costs		5%	£26,933.43
			£538,668.69

explicit approach as done previously, but this increases the complexity and usability of the method.

Summary and conclusions

As in Chapter 6 on short lease valuation, a summary of the results of the valuations produced by the various approaches is displayed in Table 8.12. As in Table 6.7, the same qualifications about the construction and interpretation of the table apply.

There is less of a clear-cut pattern in the valuations, though with one exception the traditional approaches tend to produce valuations below the level of the contemporary and advanced approaches once again. Another common factor, however, is that the more explicit approaches can again be viewed as more rational than the traditional models which, due to their implicit nature, do tend to be approximations of the actual expected pattern of value. The traditional valuations tend to be very sensitive to the assumptions made in their construction. The methods that are the least sensitive to variation in the input assumptions are the simulation and the arbitrage approach. This consistency is a very strong factor in recommending these methods for wider use.

Investment Property Valuation Today

Table 8.12 Performance by valuation approach

Valuation approach	Valuation £	Difference from mean %	s1 %	s2 %	s3 %	High comb. %	Low comb. %
Initial yield	529,100	−3.368	12.50	10.00	n.a.	n.a.	n.a.
Top slice approach	595,027	8.673	0.30	11.43	4.56	16.06	18.69
Equivalent yield term and reversion	530,920	−3.035	5.69	10.35	n.a.	17.50	15.73
Split yield term and reversion	506,237	−7.543	3.62	5.69	6.27	5.28	8.81
'Short-cut' DCF	586,318	7.082	5.92	7.08	2.66	12.79	13.06
'Full' DCF	545,788	−0.320	6.68	3.94	5.38	13.65	18.08
Simulation	548,259	0.131	0.13	6.00	3.88	10.50	8.36
Arbitrage	538,669	−1.620	3.83	1.93	0.27	2.23	8.32
Mean	**£547,540**						

Chapter 9

Conclusions

Investment valuers today have a bewildering array of possible approaches and methods available to them. In fact, there may well be too many options available. Each approach has different characteristics and requires different assumptions. Two valuers tackling the same valuation task with the same set of evidence could arrive at a different opinion of value simply due to the choice of methods they each select. Theoretically this should not happen: all methods, with the same assumptions made, should produce the same answer. The problem is that with the range of approaches available it is not obvious whether the assumptions made in different models are equivalent. The RICS has never wanted to be too prescriptive in this area but this is surely an area that needs to be looked at very closely.

The review of how the various approaches can be applied to contemporary market situations has illustrated that all of the methods can, with adaptation, be applied to any given situation and can all produce a successful (ie acceptable) valuation. Each model's performance is tempered by its inherent characteristics, the areas examined in Part 1 of this book. These characteristics will determine the degree of risk of failure to produce an acceptable valuation that the models offer.

The various traditional methods of valuation show the lowest risk of failure where they are dealing with simple investments with long secure income streams and where there is good market evidence. They each have inherent faults to a lesser or greater degree. Fundamentally, where the lease rent and the market rent differ, the models are inadequate. They 'fudge' the valuation of the various components of the investment to produce a figure that is acceptable to the market. They do not immediately reveal the assumptions made by the investor or the valuer about expected return or growth. They have great difficulty addressing complexity and where explicit assumptions about the future have to be made, largely because they are implicit approaches. It is in these latter areas, particularly where the evidence from the market is inadequate, that there is the greatest risk of failure.

These weaknesses are often outweighed by two principal strengths: familiarity and the extent to which the market uses them. There is much to be said about doing things the way that you are used to doing them. There may be more efficient and sound ways of carrying out a task but quite often the best performance will come by doing something that is familiar – there is less chance of making an error and more chance that you will recognise when an error has been made. This is an important reason why traditional methods have persisted in use in commercial practice, even though the weaknesses in the models have been recognised for some 25 years.

The strength arising out of frequency of use in the market relates to the fact that if everyone is using the same techniques as you are then the mistakes you make will be cancelled out be similar ones made by the rest of the market! This may seem to be flippant but it is actually important. The main task of valuation is to predict the price of exchange of an investment as a substitute for selling it. To do this it is best to use the methods that the market uses to determine these transaction prices and also to record the value of the investment in the portfolios of investors. The fact that these methods are flawed is immaterial if valuer and owner are coming to the same answer.

This situation should be recognised as not being sustainable. The market is getting more sophisticated and is adopting more sophisticated techniques that are producing different results to the traditional methods. There are also situations in the marketplace where the weakness of the traditional models will not protect the valuer from producing a negligently different valuation from others using the same methods. Such situations occur where the investment has short leases, break clauses or is over-rented, and where there the market does not produce enough transactions to be sure about the adjustments necessary.

The use of the traditional methods will continue but valuers must be aware of where there is this greater risk of failure. This means that valuers must increase the range of techniques that they can use. These methods may act as substitutes for the traditional methods or to supplement them in an exploration of the factors that the traditional methods have the most difficulty in dealing with.

Of the alternative methods, the contemporary methods are the most easily accessible to the practising valuer. Both the 'short-cut' DCF and the 'full' DCF methods have potential advantages and disadvantages to be used in valuation practice.

The 'short-cut' DCF approach has much to commend it. It solves many of the problems that arise with the traditional approaches. It is mathematically consistent. It is explicit as to the assumptions made regarding expected return and growth assumptions. It also has the benefit of being, largely, a technique whose components are derived from the analysis of market transactions. It should both reflect the sentiments of the market and produce results that are consistent with those produced by the best of the traditional methods.

There are, however, problems in using the 'short-cut' DCF that must be recognised. Chapter 6, which examined the valuation of short leased freehold investments, illustrated that it is relatively easy to produce a valuation that is out of step with that produced by other approaches. The logic that underlies the assumptions made within the models has to be carefully considered, particularly when growth rates and equated yield assumptions derived from one type of investment, ie the benchmark rack-rented freehold investment let on a long lease, is applied to a different kind of investment, namely a freehold investment let for only a short period. This is not an inherent fault with the technique; however, the method requires careful thought when being applied to different circumstances.

Part of the problem with the approach stems from one of its main strengths, that it relies on market comparison. To work at its best the method requires good market evidence, just as the traditional methods do. Where that evidence is scarce or is contradictory the approach tends to struggle almost to the same degree as the traditional approaches. This is not meant to associate this method with the traditional approaches – its explicit nature, stability and mathematical soundness makes it considerably superior to these methods. However, it is not the universal solution to all the problems of the valuer, and must be applied with caution.

The 'full' DCF approach has many disadvantages, particularly when compared with the 'short-cut' DCF approach. It removes the security of reference to market transactions from the valuer. It requires them to make explicit assumptions about a number of uncertain future events in the life of an investment. It cannot be expressed as a single formula that can be tested for validity. Its construction is prone to the effect of the subjective input of the valuer and, although it is usually not that sensitive to individual assumptions, it is highly sensitive to small variations in assumptions taken together.

Despite all this, the 'full' DCF approach has a lot of merit, particularly in complex investments. Its very nature makes the valuer address the key factors in the valuation. The valuer must decide a range for the most likely future of the investment and state it in the valuation. These predictions are actual forecasts rather than implied values (remember there is much in the 'short-cut' DCF that remains implicit). The fact that the assumptions made in the valuation are so transparent may well make the valuer be realistic about them. It should certainly not allow factors that are glossed over in the other approaches to be submerged in the valuation. The approach is very flexible and adaptable, and can tackle just about any problem. It can, of course, also allow a valuer to proceed where market evidence is limited.

This does not entirely outweigh the problems. There is a natural reluctance among valuers to employ a technique that is transparent, not related to market evidence and reputed to be unstable. Valuers have been weaned on stories that treat 'full' DCFs as the big bad wolf that will gobble them up if they go into the woods and play with it. This is based on the view that the traditional methods are absolutely reliable and will always produce similar figures to other competent valuers. In fact, the evidence is that even with competent valuers using the same traditional methods valuation opinions can vary markedly – more markedly perhaps than valuers expect. While the 'full' DCF approach may well not produce greater variations between valuers than the traditional methods, this is not to say that it should be employed in isolation. However, it is a valuable technique that should not be ignored.

Many of the problems with forecasting can be addressed by employing a simulation approach. The fact that this method allows valuers to express confidence in their predictions provides a new dimension to the quality of a valuation. The development of relatively inexpensive and easy to apply add-on software such as @Risk brings this technique within easy reach of most valuers. While simulation has the disadvantage of simply not being expressible as a formula, it is conceptually easy to understand and is reasonably straightforward to apply. Of the methods we have looked at in this book, it is simulation that perhaps extends the capability of the valuer to the greatest extent and has the most potential for future application.

This is not to dismiss the attractions of the other 'advanced' method we looked at – arbitrage based valuation. On the one hand, this method does have its problems for application in practice. It is a different type of approach and requires the valuer to think about concepts that have not really been on the agenda in the past. It also requires good market evidence to work really effectively. The analysis of the DCY is not as easy as the analytical requirements of some other methods, and in its implied form it does not reveal very much about the assumptions of investors. On the other hand, setting these difficulties aside, once the principles have been understood it is a very easy method to apply and is also very stable in execution – in fact it is the most stable of all the methods we have examined. This is a major benefit. However, in terms of explicitness and when dealing with specific complex investments, it remains inferior to the 'full' DCF and simulation approaches.

Some final thoughts

This is where we find ourselves at the beginning of the third millennium. We have a much more complex property investment market. We also have a range of methods we can employ, depending on the circumstances. None of the approaches are perfect, all have good and bad points. The best thing a valuer can do is understand these good and bad points, and appreciate when it is best to apply which method. A good rule to follow, though, may well be 'one method good, two or more methods better'. The nature of the property investment market is following an increasing trend towards complexity and hetereogeneity. It seems increasingly unlikely that valuers will be able to use a single approach and be confident that the value outcome is correct. The watchword for valuers in the future must be to be open to new ideas and to be flexible, but also to be cautious about their use.

References and Further Reading

Baum, A. and Crosby, N. (1991) *Property Investment Appraisal*. Routledge, London.

Baum, A. and Crosby, N. (1995) 'Over-rented properties: bond or equity? A case of market value, investment worth and actual price', *Journal of Property Valuation and Investment*, vol. 13, no. 2.

Baum, A., Mackmin, D. and Nunnington, N. (1997) *The Income Approach to Property Valuation*, 4th edn. Thompson, London.

Bowcock, P. (1983) 'The valuation of varying incomes', *Journal of Valuation*, vol. 1, pp. 366–71.

Brown, G. (1991) *Property Investment and the Capital Markets*. E&FN Spon, London.

Crosby, N. (1994) 'Discounted cash flow techniques: price or worth?', *Estates Gazette*, 4 July 1994.

Crosby, N. and Murdoch, S. (1994) 'Capital valuation implications of rent free periods', *Journal of Property Valuation and Investment*, vol. 12, pp. 51–66.

Crosby, N. and Murdoch, S. (1998) 'Changing lease structures in commercial property markets', Paper 2 of *Right Space: Right Price?*, RICS Research, March.

Draper, D. and Chapman-Findlay, M., (1982) 'Capital asset pricing and real estate valuation', *AREUEA Journal*, vol. 10, no. 2, pp. 152–83.

Epstein, D. (1993) 'The "core yield"', *Estates Gazette*, 1 May.

French, N. and Ward, C. (1995) 'Valuation and arbitrage', *Journal of Property Research*, vol. 12, pp. 1–11.

French, N. and Ward, C. (1996) 'Applications of the arbitrage method of valuation', *Journal of Property Research*, vol. 13, pp. 47–56.

French, N. and Ward, C. (1997) 'The valuation of upwards-only rent reviews: an option pricing model', *Journal of Property Valuation and Investment*, vol. 15, no. 2.

Geltner, D. (1989) 'On the use of the financial option price model to value and explain vacant urban land', *AREUEAJournal*, vol. 17, pp. 142–58.

Havard, T.M. (1999a) 'Valuer Behaviour and the Causes of Excessive Variance in Commercial Investment Property Valuation', unpublished PhD thesis, UMIST.

Havard, T.M. (1999b) *Why Do Valuers Get It Wrong? A Survey of Senior Practitioners*. RICS 'Cutting Edge' Conference, Cambridge. Available from www.rics.org.uk/research.

Herd, G. and Lizieri, C. (1994) 'Valuing and appraising new lease forms: the case of break clauses in office markets', *Proceedings, RICS 'Cutting Edge' Conference*. RICS, London, pp. 127–44.

Hutchinson, N., Adair, A., MacGregor, B., McGreal, S. and Nathakumaran, N. (1996) *Variations in the Capital Valuations of UK Commercial Property*. RICS, London.

Macfarlane, J. (1995) 'The use of simulation in investment analysis', *Journal of Property Valuation and Investment*, vol. 13, no. 4.

Mackmin, D. (1995) 'DCF discounted: further implications for the valuation surveyor arising from the over-rented property debate', *Journal of Property Valuation and Investment*, vol. 13, no. 2.

Nathakumaran, N. and MacLeary, A. R. (eds) (1998) *Property Investment Theory*. E&FN Spon, London.

Peto, R., French, N. and Bowman, G. (1996) 'Price and worth developments in valuation methodology', *Journal of Property Valuation and Investment*, vol. 14, no. 4.

RICS (1997) *Commercial Investment Property: Valuation Methods – An Information Paper*. RICS, London.

RICS (1998) *Right Space: Right Price?*, Papers 1–6. RICS, London.

Ross, S. (1976) 'The arbitrage theory of capital asset pricing', *Journal of Economic Theory*, vol. 13, pp. 341–60.

Wood, E. (1972) *Property Investment: A Real Value Approach*. PhD thesis, University of Reading.